SI Base Units

Base Quantity	Name of Unit	Symbol
Length	meter	m
Mass	kilogram	kg
Time	second	s
Electrical current	ampere	A
Temperature	kelvin	K
Amount of substance	mole	mol
Luminous intensity	candela	cd

Derived Units in the SI System

Physical Quantity	Name	Symbol	Units
Energy	Joule	J	$kg\ m^2\ s^{-2}$
Force	Newton	N	$kg\ m\ s^{-2}$
Power	Watt	W	$kg\ m^2\ s^{-3}$
Electric charge	Coulomb	C	$A\ s$
Electrical resistance	Ohm	Ω	$kg\ m^2\ s^{-3}\ A^{-2}$
Electrical potential difference	Volt	V	$kg\ m^2\ s^{-3}\ A^{-1}$
Electrical capacity	Farad	F	$A^2\ s^4\ kg^{-1}\ m^{-2}$
Frequency	Hertz	Hz	s^{-1}

Some Commonly Used Non-SI Units

Unit	Quantity	Symbol	Conversion Factor
Angstrom	Length	Å	$1\ Å = 10^{-10}\ m = 100\ pm$
Calorie	Energy	cal	$1\ cal = 4.184\ J$
Debye	Dipole moment	D	$1\ D = 3.3356 \times 10^{-30}\ C\ m$
Gauss	Magnetic field	G	$1\ G = 10^{-4}\ T$
Liter	Volume	L	$1\ L = 10^{-3}\ m^3$

Prefixes Used with SI Units

Prefix	Symbol	Meaning
Tera-	T	1,000,000,000,000, or 10^{12}
Giga-	G	1,000,000,000, or 10^9
Mega-	M	1,000,000, or 10^6
Kilo-	k	1,000, or 10^3
Deci-	d	1/10, or 10^{-1}
Centi-	c	1/100, or 10^{-2}
Milli-	m	1/1,000, or 10^{-3}
Micro-	μ	1/1,000,000, or 10^{-6}
Nano-	n	1/1,000,000,000, or 10^{-9}
Pico-	p	1/1,000,000,000,000, or 10^{-12}

Problems and Solutions

to accompany

RAYMOND CHANG

PHYSICAL CHEMISTRY
for the Biosciences

Mark D. Marshall

AMHERST COLLEGE

Helen O. Leung

AMHERST COLLEGE

University Science Books
Mill Valley, California

University Science Books
www.uscibooks.com

The cover image shows water molecules as they move through the bovine aquaporin single file. A histidine residue protruding into the pore ensures that molecules larger than water cannot enter. Peter Agre shared the Nobel Prize in Chemistry in 2003 for this work. (Image courtesy of Bing Jap.)

Text Design: Paul Anagnostopoulos
Cover Design: Mark Ong
Composition: Windfall Software
Printer & Binder: Edwards Brothers, Inc.

This book is printed on acid-free paper.

ISBN 1-891389-39-4

Library of Congress Control Number 2005900205

Printed in the United States of America 10 9 8 7 6 5 4 3

Contents

Preface

In an old joke, a New Yorker's response to a tourist's question, "How do you get to Carnegie Hall?" is "Practice, practice, practice." The same could be said for achieving success in science. The only way to really learn chemistry is to do chemistry. Our intention for this solutions manual is to encourage students of physical chemistry to work many problems—even more than they are assigned! We know that our own mastery of the subject improved by solving all of the problems posed by Raymond Chang in his textbook and that new, previously undiscovered connections were revealed to us as we discussed how to approach each one. We hope that students, too, will discuss these solutions because explaining how a question is answered is the best way of being certain that you understand the material yourself.

Indeed, we may further extend the opening metaphor. Just as there is more to music than playing or singing the right notes at right time, there is more to science than simply getting the right answer. Science is a process for recognizing important questions and how to set about seeking answers to those questions. Conversation is an important part of that process, and with our own students we would be certain to discuss the assumptions involved in arriving at a solution and to explain the intermediate steps used to determine an answer. We have tried to approximate that experience in this solutions manual by adopting a similar style; only for the simplest problems are the answers merely quoted.

We have made every effort to ensure consistency in the values of the physical constants used, both internally and with those given in the textbook. Similarly, we strived for a careful and consistent choice of units and symbols. Of course, every measurement has an associated uncertainty that must be considered when comparing results from different experiments or those of experiment with predictions from theory. It is good to practice this aspect of science, too, and we have done so by paying attention to the number of significant figures used in answering each problem. The appropriate number of significant figures is determined by the data given. So, for example, if a pressure is given as 0.729 atm, then the answer to the problem should contain three significant figures. Quantities given as "1 mole" or "5 L" are taken to be exact. Intermediate calculations are done retaining one extra digit, and the final answer is rounded to the correct number of significant figures as a last step. (This is important; rounding too early will cause a loss of significance in your calculations.) If a numerical result from a previous part of a problem is used in a subsequent part, the value with the extra digit is used. In such cases, the answer to the earlier part would look similar to "98.7654 kJ mol^{-1} = 98.765 kJ mol^{-1}," where the first value, 98.7654 kJ mol^{-1}, is used later in the problem and the "correct" answer is the second. Of course, every rule has its exceptions, and we did choose to be more flexible in problems involving exponentiation and logarithms.

Throughout the preparation of this manual, Raymond Chang provided us with extremely valuable guidance; we are grateful for the opportunity to once again work with him. Jane Ellis of University Science Books kept us calm and on task when we would begin to drift towards frenzy, and we thank her for overseeing the transformation of our default-settings LaTeX manuscript into a well-designed publication. Paul C. Anagnostopoulos of Windfall Software provided timely advice regarding the discipline of EPS files from misbehaving graphics programs and handled the most incorrigible himself.

Finally, one of us has performed at Carnegie Hall, but that is another story . . .

Mark D. Marshall
Helen O. Leung
Amherst, Massachusetts

CHAPTER 2

Properties of Gases

PROBLEMS AND SOLUTIONS

2.2 Some gases, such as NO_2 and NF_2, do not obey Boyle's law at any pressure. Explain.

Boyle's law applies only at constant n and constant T. Both NO_2 and NF_2 undergo association reactions:

$$2NO_2 \rightleftharpoons N_2O_4$$

$$2NF_2 \rightleftharpoons N_2F_4$$

so n is not constant for these gases.

2.4 Some ballpoint pens have a small hole in the main body of the pen. What is the purpose of this hole?

As the pen is used, and the ink leaves the pen, the gas inside expands to fill the volume left by the ink. As the volume increases, the pressure inside the pen decreases. The hole is needed to equalize the pressure to allow for easy flow of ink.

2.6 At STP (standard temperature and pressure), 0.280 L of a gas weighs 0.400 g. Calculate the molar mass of the gas.

In Problem 2.5, a relationship between the density, $\rho = m/V$, of an ideal gas and its molar mass was found,

$$\mathcal{M} = \frac{mRT}{VP} = \frac{(0.400 \text{ g}) \left(0.08206 \text{ L atm K}^{-1} \text{ mol}^{-1}\right) (273 \text{ K})}{(0.280 \text{ L}) (1.00 \text{ atm})} = 32.0 \text{ g mol}^{-1}$$

2.8 Calculate the density of HBr in g L^{-1} at 733 mmHg and 46°C. Assume ideal-gas behavior.

Use the expression found in Problem 2.5 for the density of an ideal gas in terms of its pressure, molar mass, and temperature.

$$\rho = \frac{P\mathcal{M}}{RT} = \frac{(733 \text{ mmHg}) \left(\frac{1 \text{ atm}}{760 \text{ mmHg}}\right) \left(80.91 \text{ g mol}^{-1}\right)}{\left(0.08206 \text{ L atm K}^{-1} \text{ mol}^{-1}\right) (273 + 46) \text{ K}} = 2.98 \text{ g L}^{-1}$$

2.10 The saturated vapor pressure of mercury is 0.0020 mmHg at 300 K and the density of air at 300 K is 1.18 g L^{-1}. **(a)** Calculate the concentration of mercury vapor in air in mol L^{-1}. **(b)** What is the number of parts per million (ppm) by mass of mercury in air?

(a) An expression for the concentration of mercury vapor can be obtained by rearranging the ideal gas equation, $PV = nRT$.

$$\text{Concentration} = \frac{n}{V} = \frac{P}{RT}$$

$$= \frac{(0.0020 \text{ mmHg}) \left(\frac{1 \text{ atm}}{760 \text{ mmHg}}\right)}{(0.08206 \text{ L atm K}^{-1} \text{ mol}^{-1})(300 \text{ K})}$$

$$= 1.07 \times 10^{-7} \text{ mol L}^{-1}$$

$$= 1.1 \times 10^{-7} \text{ mol L}^{-1}$$

(b) In 1 L,

$$n_{Hg} = 1.07 \times 10^{-7} \text{ mol}$$

$$m_{Hg} = \left(1.07 \times 10^{-7} \text{ mol}\right)\left(200.6 \text{ g mol}^{-1}\right) = 2.15 \times 10^{-5} \text{ g}$$

$$m_{air} = 1.18 \text{ g}$$

$$m_{total} = m_{air} + m_{Hg} \approx m_{air} = 1.18 \text{ g}$$

$$\text{ppm of Hg in air} = \frac{m_{Hg}}{m_{total}} \times 10^6 = \frac{2.15 \times 10^{-5} \text{ g}}{1.18 \text{ g}} \times 10^6 = 18$$

2.12 Sodium bicarbonate (NaHCO$_3$) is called baking soda because when heated, it releases carbon dioxide gas, which causes cookies, doughnuts, and bread to rise during baking. **(a)** Calculate the volume (in liters) of CO$_2$ produced by heating 5.0 g of NaHCO$_3$ at 180°C and 1.3 atm. **(b)** Ammonium bicarbonate (NH$_4$HCO$_3$) has also been used as a leavening agent. Suggest one advantage and one disadvantage of using NH$_4$HCO$_3$ instead of NaHCO$_3$ for baking.

(a) The following reaction takes place when sodium bicarbonate is heated:

$$2NaHCO_3(s) \longrightarrow Na_2CO_3(s) + H_2O(g) + CO_2(g)$$

$$n_{NaHCO_3} = \frac{5.0 \text{ g}}{84.0 \text{ g mol}^{-1}} = 5.95 \times 10^{-2} \text{ mol}$$

$$n_{CO_2} = n_{NaHCO_3} \left(\frac{1 \text{ mol CO}_2}{2 \text{ mol NaHCO}_3}\right)$$

$$= \left(5.95 \times 10^{-2} \text{ mol NaHCO}_3\right)\left(\frac{1 \text{ mol CO}_2}{2 \text{ mol NaHCO}_3}\right) = 2.98 \times 10^{-2} \text{ mol}$$

$$V_{CO_2} = \frac{n_{CO_2}RT}{P} = \frac{(2.98 \times 10^{-2} \text{ mol})(0.08206 \text{ L atm K}^{-1} \text{ mol}^{-1})(273 + 180) \text{ K}}{1.3 \text{ atm}}$$

$$= 0.85 \text{ L}$$

(b) Ammonium bicarbonate decomposes upon heating according to the following equation:

$$NH_4HCO_3(s) \longrightarrow NH_3(g) + H_2O(g) + CO_2(g)$$

The advantage in using the ammonium salt is that more gas is produced per gram of reactant. (The molar mass of NH_4HCO_3 is 79.1 g mol^{-1}, smaller than that of $NaHCO_3$.) The disadvantage is that one of the gases is ammonia. The strong odor of ammonia would not make the ammonium salt a good choice for baking.

2.14 **(a)** What volume of air at 1.0 atm and 22°C is needed to fill a 0.98-L bicycle tire to a pressure of 5.0 atm at the same temperature? (Note that 5.0 atm is the gauge pressure, which is the difference between the pressure in the tire and atmospheric pressure. Initially, the gauge pressure in the tire was 0 atm.) **(b)** What is the total pressure in the tire when the gauge reads 5.0 atm? **(c)** The tire is pumped with a hand pump full of air at 1.0 atm; compressing the gas in the cylinder adds all the air in the pump to the air in the tire. If the volume of the pump is 33% of the tire's volume, what is the gauge pressure in the tire after 3 full strokes of the pump?

(a) Enough gas (air) must be added to increase the pressure in the tire by 5.0 atm, since it starts at a gauge pressure of 0.0 atm. This is the same amount of gas whose volume is desired at ambient conditions. At constant n and T

$$P_1V_1 = P_2V_2$$

Letting the pressure and volume in the tire be P_1 and V_1, respectively, and denoting the same quantities at ambient conditions as P_2 and V_2,

$$V_2 = \frac{(5.0 \text{ atm})(0.98 \text{ L})}{1.0 \text{ atm}} = 4.9 \text{ L}$$

(b) When the gauge pressure reads 5.0 atm, the total pressure in the tire is 6.0 atm.

(c) Since the tire and the pump are at the same temperature, the amount of gas each contains at 1.0 atm is proportional to their volumes. Thus three strokes of the pump will add to the tire 99% of the amount of gas that the tire originally contains at 1.0 atm. At constant V and T,

$$\frac{P_1}{n_1} = \frac{P_2}{n_2}$$

Here, $n_2 = 1.99n_1$,

$$P_2 = \frac{(1.0 \text{ atm})}{n_1}(1.99n_1) = 1.99 \text{ atm}$$

When the total pressure in the tire is 1.99 atm, the gauge pressure reads 0.99 atm.

2.16 Nitrogen forms several gaseous oxides. One of them has a density of 1.27 g L^{-1} measured at 764 mmHg and 150°C. Write the formula of the compound.

The density of a gas is related to its molar mass as shown in Problem 2.5:

$$\mathcal{M} = \frac{\rho RT}{P} = \frac{\left(1.27 \text{ g L}^{-1}\right)\left(0.08206 \text{ L atm K}^{-1} \text{ mol}^{-1}\right)(273 + 150) \text{ K}}{(764 \text{ mmHg})\left(\frac{1 \text{ atm}}{760 \text{ mmHg}}\right)} = 43.9 \text{ g mol}^{-1}$$

Some nitrogen oxides and their molar masses are NO: 30.0 g mol^{-1}; N$_2$O: 44.0 g mol^{-1}; NO$_2$: 46.0 g mol^{-1}.

The nitrogen oxide is N$_2$O.

2.18 An ultra-high-vacuum pump can reduce the pressure of air from 1.0 atm to 1.0×10^{-12} mmHg. Calculate the number of air molecules in a liter at this pressure and 298 K. Compare your results with the number of molecules in 1.0 L at 1.0 atm and 298 K. Assume ideal-gas behavior.

When $P = 1.0 \times 10^{-12}$ mmHg:

$$n = \frac{PV}{RT} = \frac{\left(1.0 \times 10^{-12} \text{ mmHg}\right)\left(\frac{1 \text{ atm}}{760 \text{ mmHg}}\right)(1.0 \text{ L})}{\left(0.08206 \text{ L atm K}^{-1} \text{ mol}^{-1}\right)(298 \text{ K})} = 5.38 \times 10^{-17} \text{ mol}$$

$$\text{Number of molecules} = \left(5.38 \times 10^{-17} \text{ mol}\right)\left(\frac{6.022 \times 10^{23} \text{ molecules}}{1 \text{ mol}}\right)$$

$$= 3.2 \times 10^7 \text{ molecules}$$

When $P = 1.0$ atm:

$$n = \frac{PV}{RT} = \frac{(1.0 \text{ atm})(1.0 \text{ L})}{\left(0.08206 \text{ L atm K}^{-1} \text{ mol}^{-1}\right)(298 \text{ K})} = 0.0409 \text{ mol}$$

$$\text{Number of molecules} = (0.0409 \text{ mol})\left(\frac{6.022 \times 10^{23} \text{ molecules}}{1 \text{ mol}}\right) = 2.5 \times 10^{22} \text{ molecules}$$

The number of molecules present when $P = 1.0$ atm is 7.8×10^{14} times greater than when $P = 1.0 \times 10^{-12}$ mmHg.

2.20 The density of dry air at 1.00 atm and 34.4°C is 1.15 g L^{-1}. Calculate the composition of air (percent by mass) assuming that it contains only nitrogen and oxygen and behaves like an ideal gas. (*Hint*: First calculate the "molar mass" of air, then the mole fractions, and then the mass fractions of O$_2$ and N$_2$.)

This problem is similar to Problem 2.17. From the density of air, calculate its molar mass, $\mathcal{M}_{\text{air}}$ (see Problem 2.5 or 2.6), which in turn yields the mole fraction of oxygen, x_{O_2}, and the mole fraction of nitrogen, x_{N_2}. Once the mole fractions are obtained, the composition of air can be calculated.

$$\mathcal{M}_{air} = \frac{\rho_{air}RT}{P_{air}} = \frac{\left(1.15 \text{ g L}^{-1}\right)\left(0.08206 \text{ L atm K}^{-1} \text{ mol}^{-1}\right)(273.2 + 34.4) \text{ K}}{1.00 \text{ atm}} = 29.03 \text{ g mol}^{-1}$$

$$x_{O_2}\mathcal{M}_{O_2} + x_{N_2}\mathcal{M}_{N_2} = \mathcal{M}_{air} = 29.03 \text{ g mol}^{-1}$$

The sum of all mole fractions is unity, that is, $x_{O_2} + x_{N_2} = 1$, or $x_{N_2} = 1 - x_{O_2}$. Use this relation in the above equation,

$$x_{O_2}\mathcal{M}_{O_2} + \left(1 - x_{O_2}\right)\mathcal{M}_{N_2} = 29.03 \text{ g mol}^{-1}$$

$$x_{O_2}\left(32.00 \text{ g mol}^{-1}\right) + \left(1 - x_{O_2}\right)\left(28.02 \text{ g mol}^{-1}\right) = 29.03 \text{ g mol}^{-1}$$

$$32.00x_{O_2} + 28.02 - 28.02x_{O_2} = 29.03$$

$$3.98x_{O_2} = 1.01$$

$$x_{O_2} = 0.254$$

Therefore, $x_{N_2} = 1 - x_{O_2} = 1 - 0.254 = 0.746$.

In 1 mol of air, there are 0.25 mol of O_2 and 0.75 mol of N_2. The corresponding masses are therefore:

$$\text{mass of } O_2 = (0.254 \text{ mol})\left(32.00 \text{ g mol}^{-1}\right) = 8.13 \text{ g}$$

$$\text{mass of } N_2 = (0.746 \text{ mol})\left(28.02 \text{ g mol}^{-1}\right) = 20.9 \text{ g}$$

Therefore,

$$\% \text{ } O_2 \text{ by mass} = \frac{8.13 \text{ g}}{8.13 \text{ g} + 20.9 \text{ g}} \times 100\% = 28\%$$

$$\% \text{ } N_2 \text{ by mass} = 1 - \% \text{ } O_2 \text{ by mass} = 1 - 28\% = 72\%$$

2.22 Two bulbs of volumes V_A and V_B are connected by a stopcock. The number of moles of gases in the bulbs are n_A and n_B, and initially the gases are at the same pressure, P, and temperature, T. Show that the final pressure of the system, after the stopcock has been opened, equals P. Assume ideal-gas behavior.

Gas A and gas B both obey the ideal gas equation, that is, before the stopcock is open,

$$PV_A = n_A RT$$

$$PV_B = n_B RT$$

When the stopcock is open,

$$V_{total} = V_A + V_B$$

$$n_{total} = n_A + n_B$$

The total pressure is

$$P_{\text{total}} = \frac{n_{\text{total}} RT}{V_{\text{total}}} = \frac{(n_A + n_B)\, RT}{V_A + V_B} = \frac{n_A RT + n_B RT}{V_A + V_B}$$

From above, $n_A RT = P V_A$ and $n_B RT = P V_B$. Upon substitution into the expression for P_{total},

$$P_{\text{total}} = \frac{P V_A + P V_B}{V_A + V_B} = \frac{P\,(V_A + V_B)}{V_A + V_B} = P$$

2.24 A mixture containing nitrogen and hydrogen weighs 3.50 g and occupies a volume of 7.46 L at 300 K and 1.00 atm. Calculate the mass percent of these two gases. Assume ideal-gas behavior.

First calculate the total number of moles of the mixture, n_{mix}, which, together with the mass of the mixture, m_{mix}, is used to determine the number of moles of N_2 and the number of moles of H_2, and, consequently, the mass percent of these gases.

$$n_{\text{mix}} = \frac{PV}{RT} = \frac{(1.00\ \text{atm})\,(7.46\ \text{L})}{\left(0.08206\ \text{L atm K}^{-1}\ \text{mol}^{-1}\right)(300\ \text{K})} = 0.3030\ \text{mol}$$

The mass of the mixture is

$$m_{\text{mix}} = n_{N_2}\mathcal{M}_{N_2} + n_{H_2}\mathcal{M}_{H_2} = 3.50\ \text{g}$$

Because $n_{N_2} + n_{H_2} = n_{\text{mix}} = 0.303\ \text{mol}$,

$$n_{H_2} = 0.3030\ \text{mol} - n_{N_2}$$

Therefore,

$$m_{\text{mix}} = n_{N_2}\mathcal{M}_{N_2} + \left(0.3030\ \text{mol} - n_{N_2}\right)\mathcal{M}_{H_2} = 3.50\ \text{g}$$

$$n_{N_2}\left(28.02\ \text{g mol}^{-1}\right) + \left(0.3030\ \text{mol} - n_{N_2}\right)\left(2.016\ \text{g mol}^{-1}\right) = 3.50\ \text{g}$$

$$28.02 n_{N_2}\ \text{mol}^{-1} + 0.6108 - 2.016 n_{N_2}\ \text{mol}^{-1} = 3.50$$

$$26.00 n_{N_2}\ \text{mol}^{-1} = 2.889$$

$$n_{N_2} = 0.1111\ \text{mol}$$

$$\text{mass of } N_2 = n_{N_2}\left(28.02\ \text{g mol}^{-1}\right) = (0.1111\ \text{mol})\left(28.02\ \text{g mol}^{-1}\right) = 3.113\ \text{g}$$

$$\text{mass of } H_2 = 3.50\ \text{g} - 3.113\ \text{g} = 0.387\ \text{g}$$

$$\text{mass \% of } N_2 = \frac{3.113\ \text{g}}{3.50\ \text{g}} \times 100\% = 88.9\%$$

$$\text{mass \% of } H_2 = \frac{0.387\ \text{g}}{3.50\ \text{g}} \times 100\% = 11.1\%$$

2.26 Death by suffocation in a sealed container is normally caused not by oxygen deficiency but by CO_2 poisoning, which occurs at about 7% CO_2 by volume. For what length of time would it be safe to be in a sealed room $10 \times 10 \times 20$ ft? [*Source*: "Eco-Chem," J. A. Campbell, *J. Chem. Educ.* **49,** 538 (1972).]

The source of the excess CO_2 is that which is exhaled and which had as its source the O_2 that was inhaled and metabolized. Thus, to calculate how much CO_2 is added to the room, calculate how much O_2 is depleted. (This assumes a 1:1 molar ratio between CO_2 formed and O_2 used with little hydrogen oxidized to H_2O by inhaled oxygen. The ratio is actually about 1.2 O_2:1 CO_2.)

The air becomes lethal (due to CO_2) after 7% of the O_2 (to 1 sig. fig.) is removed, or

$$(10 \text{ ft})(10 \text{ ft})(20 \text{ ft}) \left(\frac{27 \text{ L}}{1 \text{ ft}^3}\right) (0.07) = 3800 \text{ L of } O_2$$

A person breathes about 0.5 L of air 12 times per minute, and the air is about 20% O_2. Thus,

$$(12 \text{ min}^{-1})(0.5 \text{ L})(0.20) = 1.2 \text{ L } O_2 \text{ min}^{-1}$$

About 30% of this inhaled O_2 is absorbed in the lungs, so that a person typically uses

$$(0.30)(1.2 \text{ L } O_2 \text{ min}^{-1}) = 0.36 \text{ L } O_2 \text{ min}^{-1}$$

For a calm, quiet person about half this amount, or 0.2 L min^{-1} would be enough, where we have rounded to 1 sig. fig. Thus, one person could last

$$\left(\frac{3800 \text{ L of } O_2}{0.2 \text{ L min}^{-1}}\right) \left(\frac{1 \text{ day}}{1440 \text{ min}}\right) = 13 \text{ days}$$

2.28 A mixture of helium and neon gases is collected over water at 28.0°C and 745 mmHg. If the partial pressure of helium is 368 mmHg, what is the partial pressure of neon? (*Note:* The vapor pressure of water at 28°C is 28.3 mmHg.)

P_{Ne} can be determined by rearranging the equation $P_{total} = P_{He} + P_{Ne} + P_{H_2O}$:

$$P_{Ne} = P_{total} - P_{He} - P_{H_2O} = 745 \text{ mmHg} - 368 \text{ mmHg} - 28.3 \text{ mmHg} = 349 \text{ mmHg}$$

2.30 A piece of sodium metal reacts completely with water as follows:

$$2Na(s) + 2H_2O(l) \longrightarrow 2NaOH(aq) + H_2(g)$$

The hydrogen gas generated is collected over water at 25.0°C. The volume of the gas is 246 mL measured at 1.00 atm. Calculate the number of grams of sodium used in the reaction. (*Note:* The vapor pressure of water at 25°C is 0.0313 atm.)

First calculate the number of moles of H_2 from the ideal gas law, from which the number of moles of Na, and therefore, the mass of Na used in the reaction can be determined.

$$P_{H_2} + P_{H_2O} = P_{total}$$

$$P_{H_2} = P_{total} - P_{H_2O} = 1.00 \text{ atm} - 0.0313 \text{ atm} = 0.969 \text{ atm}$$

$$n_{H_2} = \frac{P_{H_2}V}{RT} = \frac{(0.969 \text{ atm}) (0.246 \text{ L})}{(0.08206 \text{ L atm K}^{-1} \text{ mol}^{-1}) (273.2 + 25.0) \text{ K}} = 9.74 \times 10^{-3} \text{ mol}$$

According to the chemical equation,

$$n_{Na} = n_{H_2} \left(\frac{2 \text{ mol Na}}{1 \text{ mol H}_2} \right) = \left(9.74 \times 10^{-3} \text{ mol H}_2 \right) \left(\frac{2 \text{ mol Na}}{1 \text{ mol H}_2} \right) = 1.95 \times 10^{-2} \text{ mol}$$

The mass of Na used is

$$m_{Na} = \left(1.95 \times 10^{-2} \text{ mol} \right) \left(22.99 \text{ g mol}^{-1} \right) = 0.45 \text{ g}$$

2.32 Helium is mixed with oxygen gas for deep sea divers. Calculate the percent by volume of oxygen gas in the mixture if the diver has to submerge to a depth where the total pressure is 4.2 atm. The partial pressure of oxygen is maintained at 0.20 atm at this depth.

At constant P and T, $n \propto V$. Thus, % by volume = mol %, and mole fraction is directly related to mol %.

$$P_{O_2} = x_{O_2} P_{total}$$

$$x_{O_2} = \frac{P_{O_2}}{P_{total}} = \frac{0.20 \text{ atm}}{4.2 \text{ atm}} = 0.048$$

The mole fraction of O_2 is 0.048, or the percent by volume of $O_2 = 4.8\%$.

2.34 The partial pressure of carbon dioxide in air varies with the seasons. Would you expect the partial pressure in the Northern Hemisphere to be higher in the summer or winter? Explain.

Plant photosynthesis is a major contributor to the seasonal variation of the amount of carbon dioxide in the atmosphere. Thus, in the Northern Hemisphere the partial pressure of CO_2 is higher in the winter when less CO_2 is being utilized in photosynthesis.

2.36 Describe how you would measure, by either chemical or physical means (other than mass spectrometry), the partial pressures of a mixture of gases: **(a)** CO_2 and H_2, **(b)** He and N_2.

(a) A measurement of the total pressure of the mixture can be made at known temperature and volume. A chemical separation may then be used to measure the amount of a single component. A good choice is the reaction between CO_2 and sodium hydroxide

$$CO_2(g) + 2NaOH(aq) \longrightarrow Na_2CO_3(aq) + H_2O(l)$$

This leaves only the H_2 gas and water vapor. The partial pressure of H_2 can now be determined under the same conditions of temperature and volume after correcting for the known vapor pressure of water. Finally, the partial pressure of CO_2 is calculated from

$$P_{CO_2} = P_{total} - P_{H_2}$$

(b) In this case there is no convenient chemical means of separation, but there is a significant difference in boiling points that can be utilized. As in part **(a)** the total pressure of the mixture is first measured. The temperature is then lowered until the nitrogen liquefies (b.p. N_2: 77 K).

At this point the He is still gaseous (b.p. He: 4 K), and its pressure can be measured. Charles' Law is then used to calculate the pressure of He at the temperature of the original total pressure measurement (assuming a constant volume container). The partial pressure of N_2 is then the difference between the total pressure and helium pressure at this temperature.

2.38 A 1.00-L bulb and a 1.50-L bulb, connected by a stopcock, are filled, respectively, with argon at 0.75 atm and helium at 1.20 atm at the same temperature. Calculate the total pressure and the partial pressures of each gas after the stopcock has been opened and the mole fraction of each gas. Assume ideal-gas behavior.

At constant n and T, $P_1 V_1 = P_2 V_2$, where 1 and 2 denote the state before and after the stopcock is opened, respectively.

For Ar,

$$P_2 = \frac{P_1 V_1}{V_2} = \frac{(0.75 \text{ atm}) (1.00 \text{ L})}{2.50 \text{ L}} = 0.30 \text{ atm} = P_{Ar}$$

For He,

$$P_2 = \frac{P_1 V_1}{V_2} = \frac{(1.20 \text{ atm}) (1.50 \text{ L})}{2.50 \text{ L}} = 0.720 \text{ atm} = P_{He}$$

The total pressure is

$$P = P_{Ar} + P_{He} = 0.30 \text{ atm} + 0.720 \text{ atm} = 1.02 \text{ atm}$$

The mole fractions are

$$x_{Ar} = \frac{P_{Ar}}{P} = \frac{0.30 \text{ atm}}{1.02 \text{ atm}} = 0.29$$

$$x_{He} = 1 - x_{Ar} = 1 - 0.29 = 0.71$$

2.40 Suggest two demonstrations to show that gases do not behave ideally.

One demonstration is quite common. Namely, the condensation of a gas at low temperatures and/or high pressures to form a liquid, such as was used in Problem 2.36(b). The condensation demonstrates the existence of attractive forces between molecules. A second demonstration would be to plot the compressibility factor, $Z = P\overline{V}/RT$ vs. P. Deviations from unity show that the gas does not behave ideally.

2.42 The van der Waals constants of a gas can be obtained from its critical constants, where $a = \left(27 R^2 T_c^2 / 64 P_c\right)$ and $b = \left(R T_c / 8 P_c\right)$. Given that $T_c = 562$ K and $P_c = 48.0$ atm for benzene, calculate its a and b values.

$$a = \frac{27R^2T_c^2}{64P_c} = \frac{27\left(0.08206 \text{ L atm K}^{-1} \text{ mol}^{-1}\right)^2 (562 \text{ K})^2}{64\,(48.0 \text{ atm})} = 18.7 \text{ atm L}^2 \text{ mol}^{-2}$$

$$b = \frac{RT_c}{8P_c} = \frac{\left(0.08206 \text{ L atm K}^{-1} \text{ mol}^{-1}\right)(562 \text{ K})}{8\,(48.0 \text{ atm})} = 0.120 \text{ L mol}^{-1}$$

2.44 Without referring to a table, select from the following list the gas that has the largest value of b in the van der Waals equation: CH_4, O_2, H_2O, CCl_4, Ne.

The van der Waals constant b is related to the size of the molecule, so look for the largest molecule, which is CCl_4 in this case.

2.46 At 300 K, the virial coefficients (B) of N_2 and CH_4 are $-4.2 \text{ cm}^3 \text{ mol}^{-1}$ and $-15 \text{ cm}^3 \text{ mol}^{-1}$, respectively. Which gas behaves more ideally at this temperature?

According to the equation

$$Z = 1 + \frac{B}{V} + \cdots$$

the closer to zero the value of B, the closer is Z to unity, that is, the more ideal is the gas. According to the data given, N_2 behaves more ideally than CH_4.

2.48 Consider the virial equation $Z = 1 + B'P + C'P^2$, which describes the behavior of a gas at a certain temperature. From the following plot of Z versus P, deduce the signs of B' and C' (< 0, $= 0$, > 0).

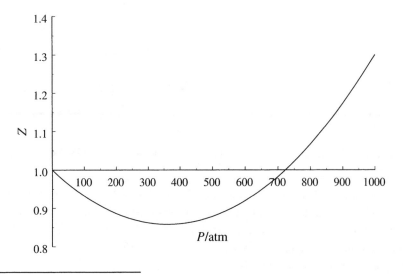

We see that the slope of the Z versus P plot is given by

$$\frac{dZ}{dP} = B' + 2C'P$$

As $P \to 0$, $\frac{dZ}{dP} \to B'$.

Additionally, the curvature of the graph is given by

$$\frac{d^2 Z}{dP^2} = 2C'$$

Near $P = 0$ the graph has a negative slope, starting at $Z = 1$ and dipping below 1. It does display positive curvature, with the graph turning around to rise above 1. Thus $B' < 0$ and $C' > 0$.

2.50 Is temperature a microscopic or macroscopic concept? Explain.

Temperature is a macroscopic concept, since one of the postulates of the kinetic theory of gases is that it deals with a very large number of molecules. Temperature is proportional to the average kinetic energy of the molecules in the system, and for this average to be meaningful, it must be taken over a large number of molecules.

2.52 If 2.0×10^{23} argon (Ar) atoms strike 4.0 cm^2 of wall per second at a 90° angle to the wall when moving with a speed of 45,000 cm s^{-1}, what pressure (in atm) do they exert on the wall?

$F = $ Force exerted by Ar atoms

$ = $ (Force exerted by 1 Ar atom) (Number of Ar atoms)

$ = $ (Change in momentum for 1 Ar atom/time) (Number of Ar atoms)

$ = \left(\dfrac{2mv}{1\,\text{s}}\right)\left(2.0 \times 10^{23}\right)$

$ = \dfrac{2\left[(39.95\ \text{amu})\left(1.661 \times 10^{-27}\ \text{kg amu}^{-1}\right)\right]\left[(45,000\ \text{cm s}^{-1})\left(\frac{1\,\text{m}}{100\,\text{cm}}\right)\right]}{1\,\text{s}}\left(2.0 \times 10^{23}\right)$

$ = 11.9\,\text{N}$

$$\frac{F}{A} = \text{Pressure exerted on the wall by Ar atoms}$$

$$= \frac{11.9\,\text{N}}{(4.0\,\text{cm}^2)\left(\frac{1\,\text{m}}{100\,\text{cm}}\right)^2}$$

$$= \left(2.98 \times 10^4\,\text{Pa}\right)\left(\frac{1\,\text{atm}}{1.01325 \times 10^5\,\text{Pa}}\right) = 0.29\,\text{atm}$$

2.54 Calculate the average translational kinetic energy for a N_2 molecule and for 1 mole of N_2 at 20°C.

For one molecule, $\overline{E}_{\text{trans}} = \frac{3}{2} k_B T$.

For one mole of molecules, $\overline{E}_{trans} = \frac{3}{2}RT$.

Therefore, for a N_2 molecule,

$$\overline{E}_{trans} = \frac{3}{2}\left(1.381 \times 10^{-23}\,\text{J K}^{-1}\right)(273+20)\;\text{K} = 6.07 \times 10^{-21}\,\text{J}$$

whereas for a mole of N_2 molecules,

$$\overline{E}_{trans} = \frac{3}{2}\left(8.314\,\text{J K}^{-1}\,\text{mol}^{-1}\right)(273+20)\;\text{K} = 3.65 \times 10^3\,\text{J mol}^{-1}$$

2.56 The c_{rms} of CH_4 is 846 m s^{-1}. What is the temperature of the gas?

The temperature of the gas can be calculated by rearranging $c_{rms} = \sqrt{\frac{3RT}{\mathcal{M}}}$

$$\frac{3RT}{\mathcal{M}} = c_{rms}^2$$

$$T = \frac{c_{rms}^2 \mathcal{M}}{3R} = \frac{\left(846\,\text{m s}^{-1}\right)^2\left(16.04 \times 10^{-3}\,\text{kg mol}^{-1}\right)}{3\left(8.314\,\text{J K}^{-1}\,\text{mol}^{-1}\right)} = 460\;\text{K}$$

Note that $v_{rms} = c_{rms}$.

2.58 At what temperature will He atoms have the same c_{rms} value as N_2 molecules at 25°C? Solve this problem without calculating the value of c_{rms} for N_2.

This problem is similar to Problem 2.55.

$$T_{He} = \frac{\mathcal{M}_{He}}{\mathcal{M}_{N_2}}T_{N_2} = \left(\frac{4.003\,\text{g mol}^{-1}}{28.02\,\text{g mol}^{-1}}\right)(273+25)\;\text{K} = 42.6\;\text{K}$$

2.60 Plot the speed distribution function for **(a)** He, O_2, and UF_6 at the same temperature, and **(b)** CO_2 at 300 K and 1000 K.

(a) In the plot, note that the heavier the molecules, the narrower the speed distribution and the smaller the most probable speed; whereas the lighter the molecules, the wider the speed distribution and the greater the most probable speed.

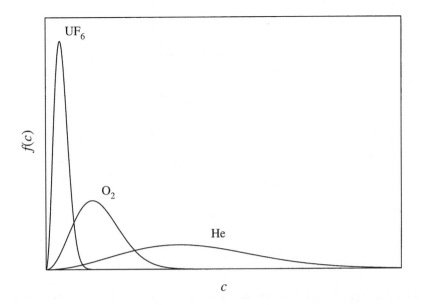

(b) In the plot, note that the lower the temperature, the narrower the speed distribution and the smaller the most probable speed; whereas the higher the temperature, the wider the speed distribution and the greater the most probable speed.

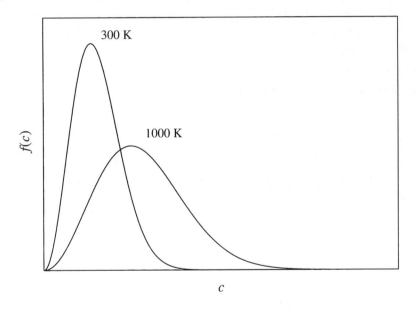

2.62 A N_2 molecule at 20°C is released at sea level to travel upward. Assuming that the temperature is constant and that the molecule does not collide with other molecules, how far would it travel (in meters) before coming to rest? Do the same calculation for a He atom. [*Hint*: To calculate the altitude, h, the molecule will travel, equate its kinetic energy with the potential energy, mgh, where m is the mass and g the acceleration due to gravity (9.81 m s^{-2}).]

This problem is an application of the law of conservation of energy. In this case, the total energy of the molecule remains the same, but it is changed from one form (kinetic energy) to another (potential energy). Therefore,

$$\text{kinetic energy} = \text{potential energy}$$

$$\frac{3}{2}k_B T = mgh$$

$$h = \frac{3k_B T}{2mg}$$

For a N_2 molecule,

$$h = \frac{3\left(1.381 \times 10^{-23}\ \text{J K}^{-1}\right)(273 + 20)\ \text{K}}{2\,(28.02\ \text{amu})\left(1.661 \times 10^{-27}\ \text{kg amu}^{-1}\right)\left(9.81\ \text{m s}^{-2}\right)} = 1.33 \times 10^4\ \text{m}$$

For a He atom,

$$h = \frac{3\left(1.381 \times 10^{-23}\ \text{J K}^{-1}\right)(273 + 20)\ \text{K}}{2\,(4.003\ \text{amu})\left(1.661 \times 10^{-27}\ \text{kg amu}^{-1}\right)\left(9.81\ \text{m s}^{-2}\right)} = 9.31 \times 10^4\ \text{m}$$

A He atom, being 7 times lighter than a N_2 molecule, would travel 7 times higher than a N_2 molecule.

2.64   At a certain temperature, the speeds of six gaseous molecules in a container are 2.0 m s^{-1}, 2.2 m s^{-1}, 2.6 m s^{-1}, 2.7 m s^{-1}, 3.3 m s^{-1}, and 3.5 m s^{-1}. Calculate the root-mean-square speed and the average speed of the molecules. These two average values are close to each other, but the root-mean-square value is always the larger of the two. Why?

The average speed of the molecules is

$$\bar{c} = \frac{\sum\limits_{i=1}^{6} c_i}{N}$$

$$= \frac{(2.0 + 2.2 + 2.6 + 2.7 + 3.3 + 3.5)\ \text{m s}^{-1}}{6}$$

$$= 2.7\ \text{m s}^{-1}$$

The mean-square speed for the molecules is

$$\overline{c^2} = \frac{\sum\limits_{i=1}^{6} c_i^2}{N}$$

$$= \frac{\left(2.0^2 + 2.2^2 + 2.6^2 + 2.7^2 + 3.3^2 + 3.5^2\right)\ \text{m}^2\ \text{s}^{-2}}{6}$$

$$= 7.67\ \text{m}^2\ \text{s}^{-2}$$

The root-mean-square speed for the molecules is

$$c_{\text{rms}} = \sqrt{\overline{c^2}} = 2.8\ \text{m s}^{-1}$$

The rms average is always larger than the straight average because squaring favors (more heavily weights) the larger values and biases the result towards the larger values.

2.66 Derive an expression for c_{mp}. [*Hint*: Differentiate $f(c)$ with respect to c in Equation 2.26 and set the result to zero.]

The most probable speed can be determined by setting $\frac{df(c)}{dc} = 0$ and then solving for c_{mp}.

$$f(c) = 4\pi c^2 \left(\frac{m}{2\pi k_B T}\right)^{\frac{3}{2}} e^{-\frac{mc^2}{2k_B T}}$$

$$= 4\pi \left(\frac{m}{2\pi k_B T}\right)^{\frac{3}{2}} c^2 e^{-\frac{mc^2}{2k_B T}}$$

In differentiating $f(c)$ with respect to c, care must be taken to apply the chain rule to the two terms c^2 and $e^{-\frac{mc^2}{2k_B T}}$.

$$\frac{df(c)}{dc} = 4\pi \left(\frac{m}{2\pi k_B T}\right)^{\frac{3}{2}} \left[2c e^{-\frac{mc^2}{2k_B T}} + c^2 e^{-\frac{mc^2}{2k_B T}} \left(-\frac{2mc}{2k_B T}\right)\right]$$

$$= 4\pi \left(\frac{m}{2\pi k_B T}\right)^{\frac{3}{2}} c e^{-\frac{mc^2}{2k_B T}} \left(2 - \frac{mc^2}{k_B T}\right)$$

When $\frac{df(c)}{dc} = 0$, $c = c_{mp}$. The above equation becomes

$$4\pi \left(\frac{m}{2\pi k_B T}\right)^{\frac{3}{2}} c_{mp} e^{-\frac{mc_{mp}^2}{2k_B T}} \left(2 - \frac{mc_{mp}^2}{k_B T}\right) = 0$$

$$2 - \frac{mc_{mp}^2}{k_B T} = 0$$

Solving for c_{mp}^2,

$$c_{mp}^2 = \frac{2k_B T}{m} \left(\frac{N_A}{N_A}\right)$$

$$= \frac{2RT}{\mathcal{M}}$$

or,

$$c_{mp} = \sqrt{\frac{2RT}{\mathcal{M}}}$$

2.68 Calculate the value of c_{mp} for C_2H_6 at 25°C. What is the ratio of the number of molecules with a speed of 989 m s^{-1} to the number of molecules with this value of c_{mp}?

$$c_{mp} = \sqrt{\frac{2RT}{\mathcal{M}}} = \sqrt{\frac{2\left(8.314 \text{ J K}^{-1} \text{mol}^{-1}\right)(273 + 25) \text{ K}}{30.07 \times 10^{-3} \text{ kg mol}^{-1}}} = 406 \text{ m s}^{-1}$$

The number of molecules with speeds between c and $c + dc$ is given by Equation 2.25,

$$\frac{dN}{N} = 4\pi c^2 \left(\frac{m}{2\pi k_B T}\right)^{\frac{3}{2}} e^{-\frac{mc^2}{2k_B T}} \, dc$$

Thus, to find the ratio of the number of molecules with speeds between c_2 and $c_2 + dc$, dN_2, to the number with speeds between c_1 and $c_1 + dc$, dN_1, (which is what is meant by the ratio of the number with speed c_2 to that with speed c_1), form the ratio $\frac{dN_2/N}{dN_1/N}$. Notice that the differential element of speed, dc, is the same for both speeds, and that it, with many other common terms, cancels

$$\frac{dN_2/N}{dN_1/N} = \frac{4\pi c_2^2 \left(\frac{m}{2\pi k_B T}\right)^{\frac{3}{2}} e^{-\frac{mc_2^2}{2k_B T}} \, dc}{4\pi c_1^2 \left(\frac{m}{2\pi k_B T}\right)^{\frac{3}{2}} e^{-\frac{mc_1^2}{2k_B T}} \, dc}$$

$$= \frac{c_2^2}{c_1^2} e^{-\frac{m}{2k_B T}\left(c_2^2 - c_1^2\right)}$$

for this problem we have $c_2 = 989$ m s^{-1} and $c_1 = 406$ m s^{-1}. Thus,

$$\frac{dN_2}{dN_1} = \left(\frac{989 \text{ m s}^{-1}}{406 \text{ m s}^{-1}}\right)^2 e^{-\left[\frac{30.07\times 10^{-3} \text{ kg mol}^{-1}}{2(8.314\text{J K}^{-1}\text{mol}^{-1})(273+25)\text{K}}\left(989^2 - 406^2\right) \text{ m}^2 \text{ s}^{-2}\right]}$$

$$= 0.0427$$

2.70 How does the mean free path of a gas depend on **(a)** the temperature at constant volume, **(b)** the density, **(c)** the pressure at constant temperature, **(d)** the volume at constant temperature, and **(e)** the size of molecules?

The mean free path is given by Equation 2.32 in the text

$$\lambda = \frac{1}{\sqrt{2}\pi d^2 \left(\frac{N}{V}\right)}$$

Although it is possible to answer this question solely by reference to the equation, it is useful to have an understanding of the physical basis for the effects observed. The key physical quantity is the density of the gas.

(a) The mean free path is independent of temperature at constant volume. T does not appear in the equation. As the temperature is increased the molecules are moving faster, but the average distance between them is not affected. The mean time between collisions decreases, but the mean distance traveled between collisions remains the same.

(b) As the density increases, the mean free path decreases, since $\frac{N}{V}$ appears in the denominator. In a more dense gas, the molecules are more closely spaced.

(c) As the pressure increases at constant temperature, the mean free path decreases. These conditions lead to a decrease in volume, hence an increase in density. The molecules are being squeezed closer together.

(d) As the volume increases at constant temperature, the mean free path increases. As the molecules move into the expanded volume, they move further apart from each other.

(e) As the size of the molecules increases, the mean free path decreases. The collision diameter, d, appears in the denominator of the equation. Larger molecules do not have to travel as far before they run into each other.

2.72 Calculate the mean free path and the binary number of collisions per liter per second between HI molecules at 300 K and 1.00 atm. The collision diameter of the HI molecules may be taken to be 5.10 Å. Assume ideal-gas behavior.

The ideal gas law is used to calculate $\frac{N}{V}$, which is then used to calculate the mean free path.

$$PV = nRT = \frac{N}{N_A} RT$$

$$\frac{N}{V} = \frac{PN_A}{RT} = \frac{(1.00 \text{ atm}) \left(6.022 \times 10^{23} \text{ mol}^{-1}\right)}{\left(0.08206 \text{ L atm K}^{-1} \text{ mol}^{-1}\right) (300 \text{ K})} = 2.446 \times 10^{22} \text{ L}^{-1} \left(\frac{1000 \text{ L}}{1 \text{ m}^3}\right)$$

$$= 2.446 \times 10^{25} \text{ m}^{-3}$$

$$\lambda = \frac{1}{\sqrt{2}\pi d^2 \left(\frac{N}{V}\right)} = \frac{1}{\sqrt{2}\pi \left(5.10 \times 10^{-10} \text{ m}\right)^2 \left(2.446 \times 10^{25} \text{ m}^{-3}\right)} = 3.53 \times 10^{-8} \text{ m}$$

The binary number of collisions depends on the average molecular speed, which is

$$\bar{c} = \sqrt{\frac{8RT}{\pi \mathcal{M}}} = \sqrt{\frac{8 \left(8.314 \text{ J K}^{-1} \text{ mol}^{-1}\right) (300 \text{ K})}{\pi \left(127.9 \times 10^{-3} \text{ kg mol}^{-1}\right)}} = 222.8 \text{ m s}^{-1}$$

$$Z_{11} = \frac{\sqrt{2}}{2}\pi d^2 \bar{c} \left(\frac{N}{V}\right)^2 = \frac{\sqrt{2}}{2}\pi \left(5.10 \times 10^{-10} \text{ m}\right)^2 \left(222.8 \text{ m s}^{-1}\right) \left(2.446 \times 10^{25} \text{ m}^{-3}\right)^2$$

$$= \left(7.702 \times 10^{34} \text{ m}^{-3} \text{ s}^{-1}\right) \left(\frac{1 \text{ m}^3}{1000 \text{ L}}\right) = 7.70 \times 10^{31} \text{ collisions L}^{-1} \text{ s}^{-1}$$

2.74 Suppose that helium atoms in a sealed container all start with the same speed, $2.74 \times 10^4 \text{ cm s}^{-1}$. The atoms are then allowed to collide with one another until the Maxwell distribution is established. What is the temperature of the gas at equilibrium? Assume that there is no heat exchange between the gas and its surroundings.

The total translational energy of the helium atoms can be determined from the initial speed of the atoms. Because energy is conserved, this is also the total translational energy of the atoms after equilibrium is reached. Translational energy is a function of temperature, thus, the latter can be calculated once the former is known.

Suppose there are N helium atoms. Because all the atoms have the same speed, the total translational energy is

$$E_{\text{trans}} = N \left(\frac{1}{2}mv^2\right) = N \left(\overline{E}_{\text{trans}}\right)$$

$$N \left(\frac{1}{2}mv^2\right) = N \left(\frac{3}{2}k_B T\right)$$

$$T = \frac{mv^2}{3k_B} = \frac{(4.003 \text{ amu})\left(1.661 \times 10^{-27} \text{ kg amu}^{-1}\right)\left(2.74 \times 10^4 \text{ cm s}^{-1}\right)^2 \left(\frac{1 \text{ m}}{100 \text{ cm}}\right)^2}{3\left(1.381 \times 10^{-23} \text{ J K}^{-1}\right)} = 12.0 \text{ K}$$

2.76   Calculate the values of Z_1 and Z_{11} for mercury (Hg) vapor at 40°C, both at $P = 1.0$ atm and at $P = 0.10$ atm. How do these two quantities depend on pressure? The collision diameter of Hg is 4.26 Å.

The number density and average molecular speed need to be determined before calculating Z_1 and Z_{11}.

At $P = 1.0$ atm,

$$\frac{N}{V} = \frac{PN_A}{RT} \text{ (See Problem 2.72)}$$

$$= \frac{(1.0 \text{ atm})\left(6.022 \times 10^{23} \text{ mol}^{-1}\right)}{\left(0.08206 \text{ L atm K}^{-1} \text{ mol}^{-1}\right)(273 + 40) \text{ K}} = 2.34 \times 10^{22} \text{ L}^{-1}\left(\frac{1000 \text{ L}}{1 \text{ m}^3}\right)$$

$$= 2.34 \times 10^{25} \text{ m}^{-3}$$

$$\bar{c} = \sqrt{\frac{8RT}{\pi \mathcal{M}}} = \sqrt{\frac{8\left(8.314 \text{ J K}^{-1} \text{ mol}^{-1}\right)(273 + 40) \text{ K}}{\pi\left(200.6 \times 10^{-3} \text{ kg}\right)}} = 181.8 \text{ m s}^{-1}$$

$$Z_1 = \sqrt{2}\pi d^2 \bar{c}\frac{N}{V} = \sqrt{2}\pi\left(4.26 \times 10^{-10} \text{ m}\right)^2\left(181.8 \text{ m s}^{-1}\right)\left(2.34 \times 10^{25} \text{ m}^{-3}\right)$$

$$= 3.4 \times 10^9 \text{ collisions s}^{-1}$$

$$Z_{11} = \frac{\sqrt{2}}{2}\pi d^2 \bar{c}\left(\frac{N}{V}\right)^2 = \frac{1}{2}Z_1\left(\frac{N}{V}\right)$$

$$= \frac{1}{2}\left(3.4 \times 10^9 \text{ collisions s}^{-1}\right)\left(2.34 \times 10^{25} \text{ m}^{-3}\right)$$

$$= 4.0 \times 10^{34} \text{ collisions m}^{-3} \text{ s}^{-1}$$

For an ideal gas, $\frac{N}{V} = \frac{PN_A}{RT}$. The results above show $Z_1 \propto P$, whereas $Z_{11} \propto P^2$. A reduction in P to one tenth its original value (from 1.0 atm to 0.10 atm) will likewise reduce Z_1 to one tenth its value at $P = 1.0$ atm, but Z_{11} will decrease to $\frac{1}{10^2} = \frac{1}{100}$ of its value at $P = 1.0$ atm. That is, at $P = 0.10$ atm,

$$Z_1 = 3.4 \times 10^8 \text{ collisions s}^{-1}$$

$$Z_{11} = 4.0 \times 10^{32} \text{ collisions m}^{-3} \text{ s}^{-1}$$

2.78   An inflammable gas is generated in marsh lands and sewage by a certain anaerobic bacterium. A pure sample of this gas was found to effuse through an orifice in 12.6 min. Under identical conditions of temperature and pressure, oxygen takes 17.8 min to effuse through the same orifice. Calculate the molar mass of the gas, and suggest what this gas might be.

$$\text{Rate of effusion (r)} \propto \frac{1}{\text{Time required for effusion (t)}}$$

Therefore,

$$\frac{t_{O_2}}{t_{gas}} = \frac{r_{gas}}{r_{O_2}} = \sqrt{\frac{\mathcal{M}_{O_2}}{\mathcal{M}_{gas}}}$$

$$\mathcal{M}_{gas} = \mathcal{M}_{O_2} \left(\frac{t_{gas}}{t_{O_2}}\right)^2 = \left(32.00 \text{ g mol}^{-1}\right) \left(\frac{12.6 \text{ min}}{17.8 \text{ min}}\right)^2 = 16.0 \text{ g mol}^{-1}$$

The gas is likely CH_4.

2.80 In 2.00 min, 29.7 mL of He effuse through a small hole. Under the same conditions of temperature and pressure, 10.0 mL of a mixture of CO and CO_2 effuse through the hole in the same amount of time. Calculate the percent composition by volume of the mixture.

$$\frac{r_{He}}{r_{mix}} = \sqrt{\frac{\mathcal{M}_{mix}}{\mathcal{M}_{He}}}$$

$$\mathcal{M}_{mix} = \left(\frac{r_{He}}{r_{mix}}\right)^2 \mathcal{M}_{He} = \left(\frac{\frac{29.7 \text{ mL}}{2.00 \text{ min}}}{\frac{10.0 \text{ mL}}{2.00 \text{ min}}}\right)^2 \left(4.003 \text{ g mol}^{-1}\right) = 35.31 \text{ g mol}^{-1}$$

Let x_{CO} be the mole fraction of CO, and x_{CO_2} be the mole fraction of CO_2. These mole fractions are related by

$$x_{CO} + x_{CO_2} = 1$$

The effective molar mass of the mixture can be obtained from the mole fractions and molar masses of its components:

$$x_{CO}\mathcal{M}_{CO} + x_{CO_2}\mathcal{M}_{CO_2} = \mathcal{M}_{mix}$$

$$x_{CO}\mathcal{M}_{CO} + \left(1 - x_{CO}\right)\mathcal{M}_{CO_2} = \mathcal{M}_{mix}$$

$$x_{CO}\left(28.01 \text{ g mol}^{-1}\right) + \left(1 - x_{CO}\right)\left(44.01 \text{ g mol}^{-1}\right) = 35.31 \text{ g mol}^{-1}$$

$$28.01x_{CO} + 44.01 - 44.01x_{CO} = 35.31$$

$$16.00x_{CO} = 8.70$$

$$x_{CO} = 0.54$$

At constant P and T, $n \propto V$. Therefore, volume fraction = mole fraction. As a result,

% of CO by volume $= 54\%$

% of CO_2 by volume $= 1 - \%$ of CO by volume $= 46\%$

2.82 An equimolar mixture of H_2 and D_2 effuses through an orifice at a certain temperature. Calculate the composition (in mole fractions) of the gas that passes through the orifice. The molar mass of deuterium is 2.014 g mol^{-1}.

The natural abundance of deuterium (D) is so small that to four significant figures the molar mass of $^1H_2$ is the same as the average molar mass of naturally occuring H_2, but for D_2 the molar mass is 4.028 g mol^{-1}.

$$\frac{r_{H_2}}{r_{D_2}} = \sqrt{\frac{\mathcal{M}_{D_2}}{\mathcal{M}_{H_2}}} = \sqrt{\frac{4.028 \text{ g mol}^{-1}}{2.016 \text{ g mol}^{-1}}} = 1.414$$

The amount of H_2 that passes through the orifice is 1.414 times more than that of D_2. Therefore,

$$\text{mole fraction of } H_2 \text{ that passes the orifice} = \frac{1.414}{1.414 + 1} = 0.5857$$

$$\text{mole fraction of } D_2 \text{ that passes the orifice} = 1 - \text{mole fraction of } H_2 \text{ that passes the orifice}$$

$$= 0.4143$$

2.84 A barometer with a cross-sectional area of 1.00 cm^2 at sea level measures a pressure of 76.0 cm of mercury. The pressure exerted by this column of mercury is equal to the pressure exerted by all the air on 1 cm^2 of Earth's surface. Given that the density of mercury is 13.6 g cm^{-3} and the average radius of Earth is 6371 km, calculate the total mass of Earth's atmosphere in kilograms. (*Hint*: The surface area of a sphere is $4\pi r^2$, where r is the radius of the sphere.)

The total mass of Earth's atmosphere can be determined from the mass of air on 1 cm^2 of Earth's surface and the surface area of the Earth.

Mass of air on 1 cm^2 of Earth's surface = Mass of 76.0 cm mercury/cm^2

$$= (76.0 \text{ cm}) \left(13.6 \text{ g cm}^{-3}\right) = 1.034 \times 10^3 \text{ g cm}^{-2}$$

Surface area of the Earth $= 4\pi r^2 = 4\pi \left(6371 \times 10^3 \text{ m}\right)^2 \left(\frac{100 \text{ cm}}{1 \text{ m}}\right)^2 = 5.1006 \times 10^{18} \text{ cm}^2$

$$\text{Mass of atmosphere} = \left(1.034 \times 10^3 \text{ g cm}^{-2}\right)\left(5.1006 \times 10^{18} \text{ cm}^2\right)$$

$$= 5.27 \times 10^{21} \text{ g} = 5.27 \times 10^{18} \text{ kg}$$

2.86 A stockroom supervisor measured the contents of a partially filled 25.0-gallon acetone drum on a day when the temperature was 18.0°C and the atmospheric pressure was 750 mmHg, and found that 15.4 gallons of the solvent remained. After tightly sealing the drum, an assistant dropped the drum while carrying it upstairs to the organic laboratory. The drum was dented and its internal volume was decreased to 20.4 gallons. What is the total pressure inside the drum after the accident? The vapor pressure of acetone at 18.0°C is 400 mmHg. (*Hint*: At the time the drum was sealed, the pressure inside the drum, which is equal to the sum of the pressures of air and acetone, was equal to the atmospheric pressure.)

First calculate the pressure of air sealed in the drum before it was dented, $P_{\text{air, i}}$. Then from Boyle's law calculate the pressure of air in the drum after it was dented, $P_{\text{air, f}}$. The partial pressure (or vapor pressure) of acetone remains constant as T is constant.

$$P_{\text{air, i}} + P_{\text{acetone}} = 750 \text{ mmHg}$$

$$P_{\text{air, i}} = 750 \text{ mmHg} - P_{\text{acetone}} = 750 \text{ mmHg} - 400 \text{ mmHg} = 350 \text{ mmHg}$$

The volumes occupied by air before and after the drum was dented are, respectively,

$$V_i = 25.0 \text{ gallons} - 15.4 \text{ gallons} = 9.6 \text{ gallons}$$

$$V_f = 20.4 \text{ gallons} - 15.4 \text{ gallons} = 5.0 \text{ gallons}$$

At constant n_{air} and T,

$$P_{air, i}V_i = P_{air, f}V_f$$

$$P_{air, f} = \frac{P_{air, i}V_i}{V_f} = \frac{(350 \text{ mmHg}) (9.6 \text{ gallons})}{5.0 \text{ gallons}} = 6.72 \times 10^2 \text{ mmHg}$$

The total pressure inside the drum after the accident is

$$P = P_{air, f} + P_{acetone} = 6.7 \times 10^2 \text{ mmHg} + 400 \text{ mmHg} = 1.07 \times 10^3 \text{ mmHg}$$

2.88 In terms of the hard-sphere gas model, molecules are assumed to possess finite volume, but there is no interaction among the molecules. **(a)** Compare the $P - V$ isotherm for an ideal gas and that for a hard-sphere gas. **(b)** Let b be the effective volume of the gas. Write an equation of state for this gas. **(c)** From this equation, derive an expression for $Z = P\overline{V}/RT$ for the hard-sphere gas and make a plot of Z versus P for two values of T (T_1 and T_2, $T_2 > T_1$). Be sure to indicate the value of the intercepts on the Z axis. **(d)** Plot Z versus T for fixed P for an ideal gas and for the hard-sphere gas.

(a) For any given pressure the volume of a hard-sphere gas will be greater than that of the ideal gas for which there is no excluded volume. This difference will be greatest at high pressure and become smaller, approaching zero as the pressure approaches zero. Furthermore, the molar volume can never get smaller than b, where b is the effective volume of a mole of the gas. See graph below.

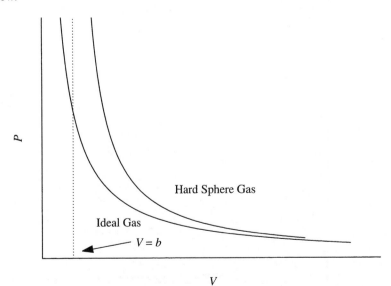

(b) The equation of state would be similar to that of the van der Waals gas, but with a value of zero for a, since there are no intermolecular attractions. Thus,

$$P\left(\overline{V} - b\right) = RT$$

(c) By definition, $Z = \frac{P\overline{V}}{RT}$, and from part **(b)**, $P\left(\overline{V} - b\right) = P\overline{V} - Pb = RT$. Thus,

$$\frac{P\overline{V}}{RT} - \frac{Pb}{RT} = 1$$

or,

$$Z = 1 + \frac{b}{RT}P$$

The graph of Z versus P is seen to be a straight line with a slope of $\frac{b}{RT}$ and as $P \to 0$, $Z \to 1$. This results in the plot below for $T_2 > T_1$.

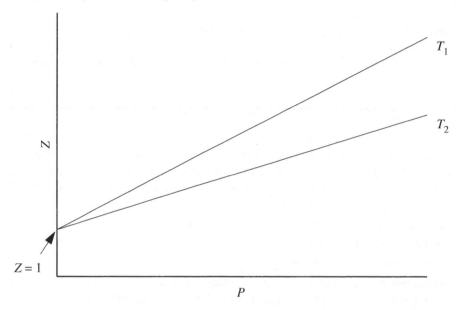

(d) For the ideal gas, $Z = 1$ for all combinations of pressure and temperature, but for the hard-sphere gas, part **(c)** shows

$$Z = 1 + \frac{Pb}{R}\frac{1}{T}$$

This is illustrated for fixed P in the plot below.

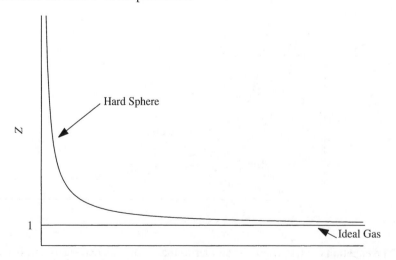

2.90 You may have witnessed a demonstration in which a burning candle standing in water is covered by an upturned glass. The candle goes out and the water rises in the glass. The explanation usually given for this phenomenon is that the oxygen in the glass is consumed by combustion, leading to a decrease in volume and hence the rise in the water level. However, the loss of oxygen is only a minor consideration. **(a)** Using $C_{12}H_{26}$ as the formula for paraffin wax, write a balanced equation for the combustion. Based on the nature of the products, show that the predicted rise in water level due to the removal of oxygen is far less than the observed change. **(b)** Devise a chemical process that would allow you to measure the volume of oxygen in the trapped air. (*Hint:* Use steel wool.) **(c)** What is the main reason for the water rising in the glass after the flame is extinguished?

(a) The combustion reaction is

$$C_{12}H_{26}(s) + \frac{37}{2}O_2(g) \longrightarrow 12CO_2(g) + 13H_2O(l)$$

Thus even though O_2 is being used up, some CO_2 gas is being formed to take its place. The mole ratio is $\frac{12}{18.5} \approx \frac{2}{3}$, which is proportional to the volumes involved. Thus, $\frac{2}{3}$ of the volume of O_2 used is replaced by CO_2. Some CO_2 does dissolve in the water, but this effect is small.

(b) A reaction that does not result in a gaseous product must be used. A suitable one is the oxidation of iron to iron oxide, or the process of rusting. If a piece of steel wool is placed in the glass, it will rust over time, using most of the O_2 in the glass. The water level will be observed to rise by about 20% of the air space originally present. (Diffusion of air into the glass is not important.)

(c) While the candle is burning the air trapped inside the glass is warmed by the flame and expands. After the flame goes out the air cools, and its pressure drops. The water level inside the glass rises due to the atmospheric pressure outside.

2.92 The Boyle temperature is the temperature at which the coefficient B is zero. Therefore, a real gas behaves like an ideal gas at this temperature. **(a)** Give a physical interpretation of this behavior. **(b)** Using your result for B for the van der Waals equation in Problem 2.91, calculate the Boyle temperature for argon, given that $a = 1.345$ atm L^2 mol^{-2} and $b = 3.22 \times 10^{-2}$ L mol^{-1}.

(a) At the Boyle temperature, the attractive forces are equal to the repulsive forces. Since the molecules do not exert any net forces on each other, the gas behaves as if it were an ideal gas and Boyle's law holds.

(b) At the Boyle temperature T_b,

$$B = b - \frac{a}{RT_b} = 0$$

$$T_b = \frac{a}{Rb} = \frac{1.345 \text{ atm L}^2 \text{ mol}^{-2}}{\left(0.08206 \text{ L atm K}^{-1} \text{ mol}^{-1}\right)\left(3.22 \times 10^{-2} \text{ L mol}^{-1}\right)} = 509 \text{ K}$$

The experimental Boyle temperature for Ar is 412 K.

2.94 The following apparatus can be used to measure atomic and molecular speed. A beam of metal atoms is directed at a rotating cylinder in a vacuum. A small opening in the cylinder allows the atoms to strike a target area. Because the cylinder is rotating, atoms traveling at different speeds will strike the target at different positions. In time, a layer of the metal will deposit on the target area, and the variation in its thickness is found to correspond to Maxwell's speed distribution. In one experiment, it is found that at 850° C, some bismuth (Bi) atoms struck the target at a point 2.80 cm from the spot directly opposite the slit. The diameter of the cylinder is 15.0 cm, and it is rotating at 130 revolutions per second. **(a)** Calculate the speed (m s^{-1}) at which the target is moving. (*Hint*: The circumference of a circle is given by $2\pi r$, where r is the radius.) **(b)** Calculate the time (in seconds) it takes for the target to travel 2.80 cm. **(c)** Determine the speed of the Bi atoms. Compare your result in **(c)** with the c_{rms} value of Bi at 850° C. Comment on the difference.

Rotating Cylinder

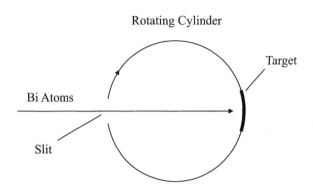

(a) The speed at which a point on the target moves is equal to the product of the circumference of the cylinder and the number of revolutions per second. Calling the speed of the target s,

$$s = \left(\frac{130\ \text{revolutions}}{1\ \text{s}}\right)\left(\frac{2\pi\left(\frac{15.0\ \text{cm}}{2}\right)}{1\ \text{revolution}}\right)\left(\frac{1\ \text{m}}{100\ \text{cm}}\right) = 61.3\ \text{m s}^{-1}$$

(b) The time, t, it takes for a point on the target to move 2.80 cm is given by

$$t = \left(\frac{2.80\ \text{cm}}{61.3\ \text{m s}^{-1}}\right)\left(\frac{1\ \text{m}}{100\ \text{cm}}\right) = 4.57 \times 10^{-4}\ \text{s}$$

(c) Since the Bi atoms struck the target at a point 2.80 cm from the spot directly opposite the slit, it took the same amount of time for the atoms to travel the 15.0 cm diameter of the cylinder as it took for the target to move 2.80 cm, as calculated in part **(b)**. Thus the speed, c, of these atoms is

$$c = \left(\frac{15.0\ \text{cm}}{4.57 \times 10^{-4}\ \text{s}}\right)\left(\frac{1\ \text{m}}{100\ \text{cm}}\right) = 328\ \text{m s}^{-1}$$

This can be compared with

$$c_{rms} = \sqrt{\frac{3RT}{\mathcal{M}}} = \sqrt{\frac{3(8.314\ \text{J K}^{-1}\ \text{mol}^{-1})(273 + 850)\ \text{K}}{209.0 \times 10^{-3}\ \text{kg mol}^{-1}}} = 366\ \text{m s}^{-1}$$

The difference is not unexpected, since c_{rms} is an average speed of the sample of atoms (assuming it is thermally equilibrated) while the speed calculated in part **(c)** is the speed of a particular group of atoms. Nevertheless, the magnitudes of the two speeds are comparable. Furthermore, the measured speed is quite close to

$$c_{mp} = \sqrt{\frac{8RT}{\pi \mathcal{M}}} = \sqrt{\frac{8(8.314 \text{ J K}^{-1}\text{ mol}^{-1})(273 + 850) \text{ K}}{\pi(209.0 \times 10^{-3}\text{ kg mol}^{-1})}} = 337 \text{ m s}^{-1}$$

which would be expected to give the most prominent spot in the measurement.

2.96   Calculate the ratio of the number of O_3 molecules with a speed of 1300 m s^{-1} at 360 K to the number with that speed at 293 K.

This problem is very similar to Problem 2.68, except that problem wanted a ratio of the numbers of molecules with two different speeds at the same temperature. Here, the ratio of the numbers of molecules with the same speed, but at two different temperatures is desired.

$$\frac{dN_2/N}{dN_1/N} = \frac{4\pi c^2 \left(\frac{m}{2\pi k_B T_2}\right)^{\frac{3}{2}} e^{-\frac{mc^2}{2k_B T_2}} dc}{4\pi c^2 \left(\frac{m}{2\pi k_B T_1}\right)^{\frac{3}{2}} e^{-\frac{mc^2}{2k_B T_1}} dc}$$

$$= \left(\frac{T_1}{T_2}\right)^{3/2} e^{-\frac{mc^2}{2k_B}\left(\frac{1}{T_2} - \frac{1}{T_1}\right)}$$

$$= \left(\frac{293 \text{ K}}{360 \text{ K}}\right)^{\frac{3}{2}} e^{-\frac{(48.00\text{ amu})(1.661\times10^{-27}\text{ kg amu}^{-1})(1300\text{ m s}^{-1})^2}{2(1.381\times10^{-23}\text{J K}^{-1})}\left(\frac{1}{360\text{ K}} - \frac{1}{293\text{ K}}\right)}$$

$$= 16.3$$

2.98   Apply your knowledge of the kinetic theory of gases to the following situations. **(a)** Two flasks of volumes V_1 and V_2 (where $V_2 > V_1$) contain the same number of helium atoms at the same temperature. **(i)** Compare the root-mean-square (rms) speeds and average kinetic energies of the helium (He) atoms in the flasks. **(ii)** Compare the frequency and the force with which the He atoms collide with the walls of their containers. **(b)** Equal numbers of He atoms are placed in two flasks of the same volume at temperatures T_1 and T_2 (where $T_2 > T_1$). **(i)** Compare the rms speeds of the atoms in the two flasks. **(ii)** Compare the frequency and the force with which the He atoms collide with the walls of their containers. **(c)** Equal numbers of He and neon (Ne) atoms are placed in two flasks of the same volume. The temperature of both gases is 74° C. Comment on the validity of the following statements: **(i)** The rms speed of He is equal to that of Ne. **(ii)** The average kinetic energies of the two gases are equal. **(iii)** The rms speed of each He atom is 1.47×10^3 m s^{-1}.

(a) With the two samples at the same temperature, (i) c_{rms} and average kinetic energies of the atoms in the two flasks are the same. Likewise, (ii) the force with which the He atoms strike the walls of the containers is the same in each flask, but the frequency of collision is greater in the smaller flask, V_1. It will have the higher pressure.

(b) Now at the same volume, but different temperatures, (i) c_{rms} is greater in the flask with the higher temperature, T_2, and (ii) in the flask with the higher temperature, T_2, the atoms collide with the walls both with greater force and with greater frequency than in the lower temperature sample.

(c) (i) The statement is false, since the lighter He atoms have a greater speed.

(c) (ii) True, since the samples are at the same temperature, the average kinetic energies of their atoms are the same.

(c) (iii) True, since

$$c_{rms} = \sqrt{\frac{3RT}{\mathcal{M}}}$$

$$= \sqrt{\frac{3(8.314 \text{ J K}^{-1} \text{ mol}^{-1})(273 + 74)\text{K}}{4.003 \times 10^{-3} \text{ kg mol}^{-1}}}$$

$$= 1.47 \times 10^3 \text{ m s}^{-1}$$

2.100   The root-mean-square velocity of a certain gaseous oxide is 493 m s^{-1} at 20° C. What is the molecular formula of the compound?

The root-mean-square velocity is used to determine the molar mass of the oxide. Given the known molar mass of oxygen, the molar mass of the other element can be determined, and therefore, its identity.

$$v_{rms} = \sqrt{\frac{3RT}{\mathcal{M}}}$$

$$\mathcal{M} = \frac{3RT}{v_{rms}^2} = \frac{3\left(8.314 \text{ J K}^{-1} \text{mol}^{-1}\right)(273 + 20) \text{ K}}{\left(493 \text{ m s}^{-1}\right)^2}$$

$$= 3.01 \times 10^{-2} \text{ kg mol}^{-1} = 30.1 \text{ g mol}^{-1}$$

Since the molar mass of the compound is less than 32 g mol^{-1}, the compound must be a monoxide. Thus, the molar mass of the other element is 30.1 g mol^{-1} – molar mass of O = 14.01 g mol^{-1}, which is the molar mass of N. The oxide is NO.

2.102   A sample of neon gas is heated from 300 K to 390 K. Calculate the percent increase in its kinetic energy.

$$\frac{\left(\overline{E}_{trans}\right)_{390}}{\left(\overline{E}_{trans}\right)_{300}} = \frac{\frac{3}{2}k_B\,(390 \text{ K})}{\frac{3}{2}k_B\,(300 \text{ K})} = 1.30$$

Therefore,

$$\% \text{ increase in kinetic energy} = 30\%$$

CHAPTER 3

The First Law of Thermodynamics

PROBLEMS AND SOLUTIONS

3.2 What is heat? How does heat differ from thermal energy? Under what condition is heat transferred from one system to another?

Heat is the transfer of energy from one object to another as a result of a temperature difference between the two objects. Heat is only transferred between systems when they are at different temperatures.

Thermal energy is that part of the energy of a system that is associated with the random motion (translational, vibrational, and rotational) of atoms and molecules.

3.4 A 7.24-g sample of ethane occupies 4.65 L at 294 K. **(a)** Calculate the work done when the gas expands isothermally against a constant external pressure of 0.500 atm until its volume is 6.87 L. **(b)** Calculate the work done if the same expansion occurs reversibly.

(a)

$$w = -P_{\text{ex}}\Delta V = -(0.500 \text{ atm})(6.87 \text{ L} - 4.65 \text{ L})\left(\frac{101.3 \text{ J}}{1 \text{ L atm}}\right) = -112 \text{ J}$$

(b)

$$w = -nRT \ln \frac{V_2}{V_1} = -\frac{7.24 \text{ g}}{30.07 \text{ g mol}^{-1}}\left(8.314 \text{ J K}^{-1} \text{ mol}^{-1}\right)(294 \text{ K}) \ln \frac{6.87 \text{ L}}{4.65 \text{ L}} = -230 \text{ J}$$

Note that work done in the reversible process in **(b)** is greater in magnitude than that in the irreversible process in **(a)**.

3.6 Calculate the work done by the reaction

$$\text{Zn}(s) + \text{H}_2\text{SO}_4(aq) \rightarrow \text{ZnSO}_4(aq) + \text{H}_2(g)$$

when 1 mole of hydrogen gas is collected at 273 K and 1.0 atm. (Neglect volume changes other than the change in gas volume.)

The gas expands until its pressure is the same as the external pressure of 1.0 atm. Furthermore, since volume changes other than gas are neglected, the change in volume of the system is

$$\Delta V = V_{H_2} = \frac{nRT}{P_{H_2}} = \frac{nRT}{P_{ex}}$$

The work done is

$$w = -P_{ex}\Delta V = -P_{ex}V_f = -P_{ex}\left(\frac{nRT}{P_{ex}}\right) = -nRT$$

$$= -(1\ \text{mol})\left(8.314\ \text{J K}^{-1}\,\text{mol}^{-1}\right)(273\ \text{K}) = -2.27 \times 10^3\ \text{J}$$

3.8 Some driver's test manuals state that the stopping distance quadruples as the velocity doubles. Justify this statement by using mechanics and thermodynamic arguments.

The kinetic energy of a moving vehicle is given by $E_{kin} = \frac{1}{2}mv^2$. If the velocity doubles, the kinetic energy quadruples. If $v_2 = 2v_1$, then $E_2 = \frac{1}{2}m\left(2v_1\right)^2 = 4E_1$. Assuming kinetic energy is dissipated at a constant rate in the brakes and between the tires and the road as heat through friction, then the stopping distance is proportional to the energy. The doubling in velocity quadruples the kinetic energy and thus the stopping distance.

3.10 An ideal gas is compressed isothermally by a force of 85 newtons acting through 0.24 meter. Calculate the values of ΔU and q.

$\Delta U = 0$ for an ideal gas undergoing an isothermal process. Using this information, q can be calculated from w, the work done on the system.

$$w = (\text{force})\,(\text{distance}) = (85\ \text{N})\,(0.24\ \text{m}) = 20\ \text{J}$$

Because $\Delta U = 0 = q + w$, $q = -w = -20$ J.

3.12 A thermos bottle containing milk is shaken vigorously. Consider the milk as the system. **(a)** Will the temperature rise as a result of the shaking? **(b)** Has heat been added to the system? **(c)** Has work been done on the system? **(d)** Has the system's internal energy changed?

The thermos bottle ensures that $q = 0$, since it prevents the transfer of heat to or from the surroundings.

(a) Energy (mechanical) has been added to the system by shaking. The random motion of the molecules in the milk is increased, and the temperature rises.

(b) The insulation of the thermos bottle prevents the transfer of heat between the system and surroundings.

(c) Work has been done on the system through shaking.

(d) The internal energy of the system has increased.

3.14   An ideal gas is compressed isothermally from 2.0 atm and 2.0 L to 4.0 atm and 1.0 L. Calculate the values of ΔU and ΔH if the process is carried out **(a)** reversibly and **(b)** irreversibly.

ΔU and ΔH of an ideal gas depend only on ΔT. Therefore, for any isothermal process [either process **(a)** or **(b)**], $\Delta U = 0$ and $\Delta H = 0$.

3.16   A piece of potassium metal is added to water in a beaker. The reaction that takes place is

$$2K(s) + 2H_2O(l) \rightarrow 2KOH(aq) + H_2(g)$$

Predict the signs of w, q, ΔU, and ΔH.

---

w is negative because the hydrogen gas produced expands and does work on the surroundings.

q is negative because thermal energy is generated in this reaction.

Since $\Delta U = q + w$, it must be negative also.

The process takes place at constant pressure, therefore $q = q_P = \Delta H$. Thus, ΔH is negative.

3.18   Consider a cyclic process involving a gas. If the pressure of the gas varies during the process but returns to the original value at the end, is it correct to write $\Delta H = q_P$?

No, it is not correct. For a cyclic process, $\Delta H = 0$, since H is a state function, but the heat, or q, associated with a process is not a state function and depends on the path. Depending on the path used to achieve the cyclic process it may take on a variety of values.

3.20   One mole of an ideal gas undergoes an isothermal expansion at 300 K from 1.00 atm to a final pressure while performing 200 J of expansion work. Calculate the final pressure of the gas if the external pressure is 0.20 atm.

$$w = -P_{ex}\left(V_2 - V_1\right) = -P_{ex}\left(\frac{nRT}{P_2} - \frac{nRT}{P_1}\right)$$

$$= -P_{ex}nRT\left(\frac{1}{P_2} - \frac{1}{P_1}\right)$$

$$\frac{1}{P_2} = -\frac{w}{P_{ex}nRT} + \frac{1}{P_1}$$

$$= -\frac{-200\ \text{J}}{(0.20\ \text{atm})\,(1\ \text{mol})\left(8.314\ \text{J K}^{-1}\,\text{mol}^{-1}\right)(300\ \text{K})} + \frac{1}{1.00\ \text{atm}} = 1.401\ \text{atm}^{-1}$$

$$P_2 = 0.71\ \text{atm}$$

3.22   A 10.0-g sheet of gold with a temperature of 18.0° C is laid flat on a sheet of iron that weighs 20.0 g and has a temperature of 55.6° C. Given that the specific heats of Au and Fe are 0.129 $J\,g^{-1}\,{}^{\circ}C^{-1}$ and 0.444 $J\,g^{-1}\,{}^{\circ}C^{-1}$, respectively, what is the final temperature of the combined metals? Assume that no heat is lost to the surroundings. (*Hint*: The heat gained by the gold must be equal to the heat lost by the iron.)

The final temperature of the sheet of gold is the same as that of the sheet of iron when thermal equilibrium is reached, and is denoted by T_f. Furthermore, the amount of heat gained by gold is the same as that lost by iron, that is,

$$q_{Au} = -q_{Fe}$$

The "−" sign is used to indicate that q_{Au} and q_{Fe} are of opposite sign.

The above relation gives

$$m_{Au}s_{Au}\Delta T_{Au} = -m_{Fe}s_{Fe}\Delta T_{Fe}$$

$$(10.0\text{ g})\left(0.129\text{ J g}^{-1}{}^{\circ}\text{C}^{-1}\right)\left(T_f - 18.0^{\circ}\text{C}\right) = -(20.0\text{ g})\left(0.444\text{ J g}^{-1}{}^{\circ}\text{C}^{-1}\right)\left(T_f - 55.6^{\circ}\text{C}\right)$$

$$\left(1.29\text{ J}{}^{\circ}\text{C}^{-1}\right)\left(T_f - 18.0^{\circ}\text{C}\right) = -\left(8.88\text{ J}{}^{\circ}\text{C}^{-1}\right)\left(T_f - 55.6^{\circ}\text{C}\right)$$

$$T_f - 18.0^{\circ}\text{C} = -\frac{8.88\text{ J}{}^{\circ}\text{C}^{-1}}{1.29\text{ J}{}^{\circ}\text{C}^{-1}}\left(T_f - 55.6^{\circ}\text{C}\right)$$

$$= -6.884\left(T_f - 55.6^{\circ}\text{C}\right)$$

$$= -6.884T_f + 382.8^{\circ}\text{C}$$

$$7.884T_f = 400.8^{\circ}\text{C}$$

$$T_f = 50.8^{\circ}\text{C}$$

3.24   The molar heat of vaporization for water is 44.01 kJ mol^{-1} at 298 K and 40.79 kJ mol^{-1} at 373 K. Give a qualitative explanation of the difference in these two values.

At the higher temperature, the water molecules have more kinetic energy and are moving faster. Thus, there are on average fewer hydrogen bonds holding the individual water molecules together. Consequently at the higher temperature, less thermal energy must be supplied to break the remaining intermolecular attractions and allow the molecules to enter the gas phase.

3.26   The heat capacity ratio (γ) of an ideal gas is 1.38. What are its $\overline{C}_P$ and $\overline{C}_V$ values?

Since $\overline{C}_P = \overline{C}_V + R$, the heat capacity ratio can be written as

$$\gamma = \frac{\overline{C}_P}{\overline{C}_V}$$

$$= \frac{\overline{C}_V + R}{\overline{C}_V}$$

This may be solved for $\overline{C}_V$ to give

$$\overline{C}_V = \frac{R}{\gamma - 1}$$

$$= \frac{R}{0.38}$$

$$= 2.6R = 22 \text{ J K}^{-1} \text{mol}^{-1}$$

and

$$\overline{C}_P = 2.6R + R = 3.6R = 30 \text{ J K}^{-1} \text{mol}^{-1}$$

3.28 Which of the following gases has the largest $\overline{C}_V$ value at 298 K? He, N_2, CCl_4, HCl.

A gas composed of non-linear, polyatomic molecules is predicted to have a larger $\overline{C}_V$ than monoatomic gases or those made up of linear molecules. In the list above, only CCl_4 is a non-linear molecule, so it should have the largest $\overline{C}_V$.

3.30 In the nineteenth century, two scientists named Dulong and Petit noticed that the product of the molar mass of a solid element and its specific heat is approximately 25 J °C^{-1}. This observation, now called Dulong and Petit's law, was used to estimate the specific heat of metals. Verify the law for aluminum (0.900 J g^{-1} °C^{-1}), copper (0.385 J g^{-1} °C^{-1}), and iron (0.444 J g^{-1} °C^{-1}). The law does not apply to one of the metals. Which one is it? Why?

Al: $(26.98 \text{ g}) \left(0.900 \text{ J g}^{-1\circ} \text{C}^{-1}\right) = 24.3 \text{ J} \circ \text{C}^{-1}$

Cu: $(63.55 \text{ g}) \left(0.385 \text{ J g}^{-1\circ} \text{C}^{-1}\right) = 24.5 \text{ J} \circ \text{C}^{-1}$

Fe: $(55.85 \text{ g}) \left(0.444 \text{ J g}^{-1\circ} \text{C}^{-1}\right) = 24.8 \text{ J} \circ \text{C}^{-1}$

Therefore, Dulong and Petit's law applies to Al, Cu, and Fe. However, since this law applies only to solids, it does not apply to mercury, a liquid. In fact, the product of the molar mass of Hg and its specific heat is

$$(200.6 \text{ g}) \left(0.139 \text{ J g}^{-1\circ} \text{C}^{-1}\right) = 27.9 \text{ J} \circ \text{C}^{-1}$$

3.32 The equation of state for a certain gas is given by $P\left[(V/n) - b\right] = RT$. Obtain an expression for the maximum work done by the gas in a reversible isothermal expansion from V_1 to V_2.

Write P in terms of n, V, and T:

$$P = \frac{RT}{\frac{V}{n} - b} = \frac{nRT}{V - nb}$$

The maximum work done by the gas undergoing an isothermal expansion is

$$w = -\int_{V_1}^{V_2} P \, dV = -\int_{V_1}^{V_2} \frac{nRT}{V - nb} \, dV$$

$$= -nRT \ln (V - nb)\big|_{V_1}^{V_2}$$

$$= -nRT \ln \frac{V_2 - nb}{V_1 - nb}$$

3.34 A quantity of 0.27 mole of neon is confined in a container at 2.50 atm and 298 K and then allowed to expand adiabatically under two different conditions: **(a)** reversibly to 1.00 atm and **(b)** against a constant pressure of 1.00 atm. Calculate the final temperature in each case.

(a) For a reversible adiabatic process, P and V are related as follow:

$$\frac{P_2}{P_1} = \left(\frac{V_1}{V_2}\right)^\gamma$$

Assuming neon to be an ideal gas, write V in terms of n, P, and T:

$$\frac{P_2}{P_1} = \left(\frac{\frac{nRT_1}{P_1}}{\frac{nRT_2}{P_2}}\right)^\gamma = \left(\frac{T_1}{T_2}\right)^\gamma \left(\frac{P_2}{P_1}\right)^\gamma$$

Rearrange the equation to give an expression relating T and P:

$$\left(\frac{T_2}{T_1}\right)^\gamma = \left(\frac{P_2}{P_1}\right)^{\gamma - 1}$$

$$\frac{T_2}{T_1} = \left(\frac{P_2}{P_1}\right)^{(\gamma - 1)/\gamma}$$

Utilizing the last equation and the fact that γ for an ideal monatomic gas is $\frac{5}{3}$, the final temperature can be calculated:

$$T_2 = T_1 \left(\frac{P_2}{P_1}\right)^{(\gamma - 1)/\gamma} = (298 \text{ K}) \left(\frac{1.00 \text{ atm}}{2.50 \text{ atm}}\right)^{\left(\frac{5}{3} - 1\right)/\frac{5}{3}} = 207 \text{ K}$$

(b) The process is adiabatic with $q = 0$. Thus, $\Delta U = w$. Since $\Delta U = C_V (T_2 - T_1)$, and $w = -P_{ex} (V_2 - V_1)$,

$$C_V (T_2 - T_1) = -P_{ex} (V_2 - V_1)$$

$$\frac{3}{2} nR (T_2 - T_1) = -P_{ex} \left(\frac{nRT_2}{P_2} - \frac{nRT_1}{P_1}\right)$$

$$\frac{3}{2} (T_2 - T_1) = -P_{ex} \left(\frac{T_2}{P_2} - \frac{T_1}{P_1}\right)$$

The expansion occurs until $P_2 = P_{ex}$. Therefore, the above expression simplifies to

$$\frac{3}{2}(T_2 - T_1) = -T_2 + \frac{P_{ex}T_1}{P_1}$$

$$\frac{5}{2}T_2 = \frac{P_{ex}T_1}{P_1} + \frac{3}{2}T_1 = \left(\frac{P_{ex}}{P_1} + \frac{3}{2}\right)T_1$$

$$T_2 = \frac{2}{5}\left(\frac{P_{ex}}{P_1} + \frac{3}{2}\right)T_1$$

$$= \frac{2}{5}\left(\frac{1.00\ atm}{2.50\ atm} + \frac{3}{2}\right)(298\ K) = 226\ K$$

3.36 A 0.1375-g sample of magnesium is burned in a constant-volume bomb calorimeter that has a heat capacity of 1769 J $^\circ$C^{-1}. The calorimeter contains exactly 300 g of water, and the temperature increases by 1.126° C. Calculate the heat given off by the burning magnesium, in kJ g^{-1} and in kJ mol^{-1}.

Heat given off by the burning magnesium is absorbed by the calorimeter and water. The heat absorbed is calculated using the heat capacity of the calorimeter, specific heat of water, and the temperature rise.

$$\text{Heat gained by calorimeter and water} = \left(1769\ J\ ^\circ C^{-1}\right)(1.126\ ^\circ C)$$

$$+ (300\ g)\left(4.184\ J\ g^{-1}\ ^\circ C^{-1}\right)(1.126\ ^\circ C)$$

$$= 3.405 \times 10^3\ J$$

Therefore, heat given off by the magnesium is

$$\frac{3.405 \times 10^3\ J}{0.1375\ g} = 2.48 \times 10^4\ J\ g^{-1} = 24.8\ kJ\ g^{-1}$$

or

$$\left(2.48 \times 10^4\ J\ g^{-1}\right)\left(24.31\ g\ mol^{-1}\right) = 6.03 \times 10^5\ J\ mol^{-1} = 603\ kJ\ mol^{-1}$$

3.38 A quantity of 2.00×10^2 mL of 0.862 M HCl is mixed with 2.00×10^2 mL of 0.431 M Ba(OH)$_2$ in a constant-pressure calorimeter that has a heat capacity of 453 J $^\circ$C^{-1}. The initial temperature of the HCl and Ba(OH)$_2$ solutions is the same at 20.48° C. For the process

$$H^+(aq) + OH^-(aq) \rightarrow H_2O(l)$$

the heat of neutralization is -56.2 kJ mol^{-1}. What is the final temperature of the mixed solution?

The final temperature of the solution can be determined once the heat capacity of the calorimeter *plus* solution is known as well as the number of moles of reaction that occurs.

First determine the number of moles of reactants.

$$n_{H^+} = \left(2.00 \times 10^{-1}\,L\right)\left(0.862\,mol\,L^{-1}\right) = 0.1724\,mol$$

$$n_{OH^-} = 2\left(2.00 \times 10^{-1}\,L\right)\left(0.431\,mol\,L^{-1}\right) = 0.1724\,mol$$

There is just enough of each reactant to completely react with the other to form 0.1724 mol of the product, H_2O. The thermal energy released by this reaction, under constant pressure, is

$$q_P = \left(-56.2\,kJ\,mol^{-1}\right)(0.1724\,mol) = -9.689\,kJ$$

This same thermal energy is used to increase the temperature of the calorimeter and its contents, but since these items are *gaining* thermal energy, the sign is switched.

$$9.689\,kJ = \left(C_{calorimeter} + m_{solution}s_{solution}\right)\Delta T$$

Two assumptions are now made. Namely, that the densities of the solutions are the same as pure water, and likewise that the specific heat of the solution is the same as that of water. (These assumptions are good to 3-5%.) Under these assumptions, $s_{solution} = 4.184\,J\,g^{-1}\,^{\circ}C^{-1}$, and

$$m_{solution} = m_{HCl} + m_{Ba(OH)_2}$$

$$= \left(2.00 \times 10^2\,mL\right)\left(1.00\,g\,mL^{-1}\right) + \left(2.00 \times 10^2\,mL\right)\left(1.00\,g\,mL^{-1}\right)$$

$$= 4.00 \times 10^2\,g$$

This allows the determination of ΔT,

$$9.689\,kJ = \left[453\,J\,^{\circ}C^{-1} + \left(4.00 \times 10^2\,g\right)\left(4.184\,J\,g^{-1}\,^{\circ}C^{-1}\right)\right]\left(\frac{1\,kJ}{1000\,J}\right)\Delta T$$

$$9.689 = \left(2.127\,^{\circ}C^{-1}\right)\Delta T$$

$$\Delta T_2 = 4.555^{\circ}\,C$$

The final temperature of the mixed solution is found from $\Delta T = T_f - T_i$, or $4.555^{\circ}\,C = T_f - 20.48^{\circ}\,C$ which gives $T_f = 25.04^{\circ}\,C$, although the assumptions made limit the accuracy of our answer to $T_f = 25.0^{\circ}\,C$.

3.40   Consider the following reaction:

$$2CH_3OH(l) + 3O_2(g) \rightarrow 4H_2O(l) + 2CO_2(g) \qquad \Delta_r H^{\circ} = -1452.8\,kJ\,mol^{-1}$$

What is the value of $\Delta_r H^{\circ}$ if **(a)** the equation is multiplied throughout by 2, **(b)** the direction of the reaction is reversed so that the products become the reactants and vice versa, and **(c)** water vapor instead of liquid water is the product?

(a) $\Delta_r H^{\circ} = 2\left(-1452.8\,kJ\,mol^{-1}\right) = -2905.6\,kJ\,mol^{-1}$

(b) $\Delta_r H^{\circ} = -\left(-1452.8\,kJ\,mol^{-1}\right) = 1452.8\,kJ\,mol^{-1}$

(c) The reaction

$$2CH_3OH(l) + 3O_2(g) \rightarrow 4H_2O(g) + 2CO_2(g)$$

is a sum of the following equations:

$$2CH_3OH(l) + 3O_2(g) \rightarrow 4H_2O(l) + 2CO_2(g) \quad \Delta_r H_1^o = -1452.8 \text{ kJ mol}^{-1}$$

$$4H_2O(l) \rightarrow 4H_2O(g) \quad \Delta_r H_2^o$$

The standard enthalpy of reaction for vaporization of H_2O is

$$\Delta_r H_2^o = 4\Delta_f \overline{H}^o [H_2O(g)] - 4\Delta_f \overline{H}^o [H_2O(l)]$$

$$= 4\left(-241.8 \text{ kJ mol}^{-1}\right) - 4\left(-285.8 \text{ kJ mol}^{-1}\right)$$

$$= 176.0 \text{ kJ mol}^{-1}$$

The standard enthalpy of reaction of $2CH_3OH(l) + 3O_2(g) \rightarrow 4H_2O(g) + 2CO_2(g)$ is

$$\Delta_r H^o = \Delta_r H_1^o + \Delta_r H_2^o = -1452.8 \text{ kJ mol}^{-1} + 176.0 \text{ kJ mol}^{-1} = -1276.8 \text{ kJ mol}^{-1}$$

3.42 The standard enthalpies of formation of ions in aqueous solution are obtained by arbitrarily assigning a value of zero to H^+ ions; that is, $\Delta_f \overline{H}^o [H^+(aq)] = 0$. **(a)** For the following reaction,

$$HCl(g) \rightarrow H^+(aq) + Cl^-(aq) \quad \Delta_r H^o = -74.9 \text{ kJ mol}^{-1}$$

calculate the value of $\Delta_f \overline{H}^o$ for the Cl^- ions. **(b)** The standard enthalpy of neutralization between a HCl solution and a NaOH solution is found to be $-56.2 \text{ kJ mol}^{-1}$. Calculate the standard enthalpy of formation of the hydroxide ion at 25° C.

(a)

$$\Delta_r H^o = -74.9 \text{ kJ mol}^{-1} = \Delta_f \overline{H}^o [H^+(aq)] + \Delta_f \overline{H}^o [Cl^-(aq)] - \Delta_f \overline{H}^o [HCl(g)]$$

$$\Delta_f \overline{H}^o [Cl^-(aq)] = -74.9 \text{ kJ mol}^{-1} - \Delta_f \overline{H}^o [H^+(aq)] + \Delta_f \overline{H}^o [HCl(g)]$$

$$= -74.9 \text{ kJ mol}^{-1} - 0 \text{ kJ mol}^{-1} + \left(-92.3 \text{ kJ mol}^{-1}\right)$$

$$= -167.2 \text{ kJ mol}^{-1}$$

(b) The neutralization reaction for 1 mole of H_2O is

$$H^+(aq) + OH^-(aq) \rightarrow H_2O(l) \quad \Delta_r H^o = -56.2 \text{ kJ mol}^{-1}$$

$$\Delta_r H^o = -56.2 \text{ kJ mol}^{-1} = \Delta_f \overline{H}^o [H_2O(l)] - \Delta_f \overline{H}^o [H^+(aq)] - \Delta_f \overline{H}^o [OH^-(aq)]$$

$$\Delta_f \overline{H}^o [OH^-(aq)] = \Delta_f \overline{H}^o [H_2O(l)] - \Delta_f \overline{H}^o [H^+(aq)] + 56.2 \text{ kJ mol}^{-1}$$

$$= -285.8 \text{ kJ mol}^{-1} - 0 \text{ kJ mol}^{-1} + 56.2 \text{ kJ mol}^{-1}$$

$$= -229.6 \text{ kJ mol}^{-1}$$

3.44 When 2.00 g of hydrazine decomposed under constant-pressure conditions, 7.00 kJ of heat were transferred to the surroundings:

$$3N_2H_4(l) \rightarrow 4NH_3(g) + N_2(g)$$

What is the $\Delta_r H^\circ$ value for the reaction?

The reaction describes the decomposition of 3 moles of hydrazine. Therefore, the amount of heat given must be scaled to this amount of reactant.

$$\Delta_r H^\circ = q_P = \left(\frac{-7.00 \text{ kJ}}{\frac{2.00 \text{ g } N_2H_4}{32.05 \text{ g mol}^{-1} N_2H_4}} \right) (3.00) = -337 \text{ kJ mol}^{-1}$$

3.46 The standard enthalpies of combustion of fumaric acid and maleic acid (to form carbon dioxide and water) are $-1336.0 \text{ kJ mol}^{-1}$ and $-1359.2 \text{ kJ mol}^{-1}$, respectively. Calculate the enthalpy of the following isomerization process:

maleic acid   fumaric acid

The chemical equations and the standard enthalpies of combustion of 1 mole of fumaric acid and 1 mole of maleic acid are given below:

$$\text{fumaric} + 3O_2 \rightarrow 4CO_2 + 2H_2O \quad \Delta_r H^\circ = -1336.0 \text{ kJ mol}^{-1}$$

$$\text{maleic} + 3O_2 \rightarrow 4CO_2 + 2H_2O \quad \Delta_r H^\circ = -1359.2 \text{ kJ mol}^{-1}$$

The isomerization reaction (maleic acid $\rightarrow$ fumaric acid) can be obtained as a combination of these two reactions:

$$4CO_2 + 2H_2O \rightarrow \text{fumaric} + 3O_2 \quad \Delta_r H^\circ = 1336.0 \text{ kJ mol}^{-1}$$

$$\text{maleic} + 3O_2 \rightarrow 4CO_2 + 2H_2O \quad \Delta_r H^\circ = -1359.2 \text{ kJ mol}^{-1}$$

Therefore, the enthalpy of the isomerization process is

$$\Delta_r H = 1336.0 \text{ kJ mol}^{-1} - 1359.2 \text{ kJ mol}^{-1} = -23.2 \text{ kJ mol}^{-1}$$

3.48 The standard molar enthalpy of formation of molecular oxygen at 298 K is zero. What is its value at 315 K? (*Hint*: Look up the $\overline{C}_P$ value in Appendix 2.)

Let $\Delta \overline{H}_1$ and $\Delta \overline{H}_2$ be the enthalpies of formation of molecular oxygen at 298 K and 315 K, respectively.

$$\Delta \overline{H}_2 = \Delta \overline{H}_1 + \overline{C}_P \Delta T = 0 \text{ J mol}^{-1} + \left(29.4 \text{ J K}^{-1} \text{ mol}^{-1}\right)(315 \text{ K} - 298 \text{ K}) = 500 \text{ J mol}^{-1}$$

3.50 The hydrogenation for ethylene is

$$C_2H_4(g) + H_2(g) \rightarrow C_2H_6(g)$$

Calculate the change in the enthalpy of hydrogenation from 298 K to 398 K. The $\overline{C}_P^0$ values are: C_2H_4: 43.6 J K^{-1} mol^{-1} and C_2H_6: 52.7 J K^{-1} mol^{-1}.

Assume the heat capacities are temperature independent.

$$\Delta_r H_{398} - \Delta_r H_{298} = \Delta C_P^0 (398 \text{ K} - 298 \text{ K})$$

$$= \left\{ \overline{C}_P^0 \left[C_2H_6(g)\right] - \overline{C}_P^0 \left[C_2H_4(g)\right] - \overline{C}_P^0 \left[H_2(g)\right] \right\} (10 \text{ K})$$

$$= \left(52.7 \text{ J K}^{-1} \text{ mol}^{-1} - 43.6 \text{ J K}^{-1} \text{ mol}^{-1} - 28.8 \text{ J K}^{-1} \text{ mol}^{-1}\right)(10 \text{ K})$$

$$= -197 \text{ kJ mol}^{-1}$$

3.52 Calculate the standard enthalpy of formation for diamond, given that

$$C(\text{graphite}) + O_2(g) \rightarrow CO_2(g) \quad \Delta_r H^0 = -393.5 \text{ kJ mol}^{-1}$$

$$C(\text{diamond}) + O_2(g) \rightarrow CO_2(g) \quad \Delta_r H^0 = -395.4 \text{ kJ mol}^{-1}$$

The formation reaction of diamond is

$$C(\text{graphite}) \rightarrow C(\text{diamond}),$$

which can be thought of as a sum of the reactions:

$$C(\text{graphite}) + O_2(g) \rightarrow CO_2(g) \qquad \Delta_r H^0 = -393.5 \text{ kJ mol}^{-1}$$

$$CO_2(g) \rightarrow C(\text{diamond}) + O_2(g) \quad \Delta_r H^0 = 395.4 \text{ kJ mol}^{-1}$$

Therefore, the standard enthalpy of formation for diamond is the sum of the standard enthalpies of reaction of the two reactions above:

$$\Delta_f H^0 (\text{diamond}) = -393.5 \text{ kJ mol}^{-1} + 395.4 \text{ kJ mol}^{-1} = 1.9 \text{ kJ mol}^{-1}$$

3.54 From the following heats of combustion,

$$CH_3OH(l) + \frac{3}{2}O_2(g) \rightarrow CO_2(g) + 2H_2O(l) \quad \Delta_r H^0 = -726.4 \text{ kJ mol}^{-1}$$

$$C(\text{graphite}) + O_2(g) \rightarrow CO_2(g) \qquad\qquad \Delta_r H^0 = -393.5 \text{ kJ mol}^{-1}$$

$$H_2(g) + \frac{1}{2}O_2(g) \rightarrow H_2O(l) \qquad\qquad\quad \Delta_r H^0 = -285.8 \text{ kJ mol}^{-1}$$

calculate the enthalpy of formation of methanol (CH_3OH) from its elements:

$$C(graphite) + 2H_2(g) + \frac{1}{2}O_2(g) \rightarrow CH_3OH(l)$$

The formation reaction of methanol can be thought of as a sum of the reactions:

$$CO_2(g) + 2H_2O(l) \rightarrow CH_3OH(l) + \frac{3}{2}O_2(g) \quad \Delta_r H^\circ = 726.4 \text{ kJ mol}^{-1}$$

$$C(graphite) + O_2(g) \rightarrow CO_2(g) \qquad\qquad \Delta_r H^\circ = -393.5 \text{ kJ mol}^{-1}$$

$$2H_2(g) + O_2(g) \rightarrow 2H_2O(l) \qquad\qquad \Delta_r H^\circ = 2\left(-285.8 \text{ kJ mol}^{-1}\right) = -571.6 \text{ kJ mol}^{-1}$$

Therefore, the standard enthalpy of formation of methanol is the sum of the standard enthalpies of reaction of the three reactions above:

$$\Delta_f H^\circ \left[CH_3OH(l)\right] = 726.4 \text{ kJ mol}^{-1} - 393.5 \text{ kJ mol}^{-1} - 571.6 \text{ kJ mol}^{-1} = -238.7 \text{ kJ mol}^{-1}$$

3.56 Calculate the difference between the values of $\Delta_r H^\circ$ and $\Delta_r U^\circ$ for the oxidation of α-D-glucose at 298 K:

$$C_6H_{12}O_6(s) + 6O_2(g) \rightarrow 6CO_2(g) + 6H_2O(l)$$

$\Delta_r H^\circ$ and $\Delta_r U^\circ$ differ from each other if the number of moles of gases after the reaction is not the same as that before the reaction.

$$\Delta_r H^\circ = \Delta_r U^\circ + RT\,\Delta n$$

Since $\Delta n = 0$, $\Delta_r H^\circ = \Delta_r U^\circ$, or $\Delta_r H^\circ - \Delta_r U^\circ = 0$.

3.58 **(a)** Explain why the bond enthalpy of a molecule is always defined in terms of a gas-phase reaction. **(b)** The bond dissociation enthalpy of F_2 is 150.6 kJ mol^{-1}. Calculate the value of $\Delta_f \overline{H}^\circ$ for $F(g)$.

(a) In the gas phase, molecules are far apart and not affected by intermolecular interactions. The bond dissociation enthalpies so determined thus refer only to the chemical bond between specific atoms and are not influenced by intermolecular interactions.

(b) It is given that $\Delta_r H^\circ = 150.6$ kJ mol^{-1} for the reaction $F_2(g) \rightarrow 2F(g)$. $\Delta_r H^\circ$ is related to the enthalpies of formation of $F_2(g)$ (which is 0) and $F(g)$ in the following manner:

$$\Delta_r H^\circ = 2\Delta_f \overline{H}^\circ \left[F(g)\right] - \Delta_f \overline{H}^\circ \left[F_2(g)\right] = 2\Delta_f \overline{H}^\circ \left[F(g)\right]$$

Therefore,

$$\Delta_f \overline{H}^\circ \left[F(g)\right] = \frac{1}{2}\Delta_r H^\circ = \frac{1}{2}\left(150.6 \text{ kJ mol}^{-1}\right) = 75.3 \text{ kJ mol}^{-1}$$

3.60 Use the bond enthalpy values in Table 3.4 to calculate the enthalpy of combustion for ethane,

$$2C_2H_6(g) + 7O_2(g) \rightarrow 4CO_2(g) + 6H_2O(l)$$

Compare your result with that calculated from the enthalpy of formation values of the products and reactants listed in Appendix 2.

Calculation of the enthalpy of combustion using bond enthalpies:

Type of bonds broken	Number of bonds broken	Bond enthalpy / kJ·mol^{-1}	Enthalpy change / kJ·mol^{-1}
C—H	12	414	4968
C—C	2	347	694
O=O	7	498.8	3491.6

Type of bonds broken	Number of bonds broken	Bond enthalpy / kJ·mol^{-1}	Enthalpy change / kJ·mol^{-1}
C=O	8	799	6392
O—H	12	460	5520

$$\Delta_r H^\circ = (4968 + 694 + 3491.6) \text{ kJ mol}^{-1} - (6392 + 5520) \text{ kJ mol}^{-1} = -2758 \text{ kJ mol}^{-1}$$

Calculation of the enthalpy of combustion using enthalpies of formation:

$$\Delta_r H^\circ = 4\Delta_f \overline{H}^\circ \left[CO_2(g)\right] + 6\Delta_f \overline{H}^\circ \left[H_2O(l)\right] - 2\Delta_f \overline{H}^\circ \left[C_2H_6(g)\right] - 7\Delta_f \overline{H}^\circ \left[O_2(g)\right]$$

$$= 4\left(-393.5 \text{ kJ mol}^{-1}\right) + 6\left(-285.8 \text{ kJ mol}^{-1}\right) - 2\left(-84.7 \text{ kJ mol}^{-1}\right) - 7\left(0 \text{ kJ mol}^{-1}\right)$$

$$= -3119.4 \text{ kJ mol}^{-1}$$

The value of $\Delta_r H^\circ$ so calculated is 13% greater than that calculated using bond enthalpies. The value determined using enthalpies of formation is the correct value, since it relies on the first law of thermodynamics. Bond enthalpies are averages determined for similar bonds in many molecules and provide estimates that are typically within 10% of the experimental value for any given, particular reaction.

3.62 Predict whether the values of q, w, ΔU, ΔH are positive, zero, or negative for each of the following processes: **(a)** melting of ice at 1 atm and 273 K, **(b)** melting of solid cyclohexane at 1 atm and the normal melting point, **(c)** reversible isothermal expansion of an ideal gas, and **(d)** reversible adiabatic expansion of an ideal gas.

For most substances, the volume increases upon melting, but for ice melting to water under the conditions given, the volume decreases. In all cases, the volume change between the solid and liquid phases is small as is the value of the work associated with the melting. For ideal gases, both internal energy, U, and enthalpy, H, depend only on temperature. Keeping these points in mind results in the following predictions.

	q	w	ΔU	ΔH
(a)	positive	positive	positive	positive
(b)	positive	negative	positive	positive
(c)	positive	negative	zero	zero
(d)	zero	negative	negative	negative

3.64   The convention of arbitrarily assigning a zero enthalpy value to all the (most stable) elements in the standard state and (usually) 298 K is a convenient way of dealing with the enthalpy changes of chemical processes. This convention does not apply to one kind of process, however. What process is it? Why?

In a nuclear process, there are different elements on both sides of the chemical equation, and this convention would not apply.

3.66   The fuel value of hamburger is approximately 3.6 kcal g^{-1}. If a person eats 1 pound of hamburger for lunch and if none of the energy is stored in his body, estimate the amount of water that would have to be lost in perspiration to keep his body temperature constant. (*Hint:* 1 lb = 454 g.)

The fuel value of 1 pound of hamburger is

$$(1 \text{ lb}) \left(\frac{454 \text{ g}}{1 \text{ lb}} \right) \left(3.6 \text{ kcal g}^{-1} \right) \left(\frac{4.184 \text{ kJ}}{1 \text{ kcal}} \right) = 6.84 \times 10^3 \text{ kJ}$$

For the vaporization of water at 298 K, $H_2O(l) \longrightarrow H_2O(g)$, $\Delta_{vap}H = 44.01$ kJ mol^{-1}. ($\Delta_{vap}H$ is appropriate, since the vaporization is taking place at constant, atmospheric pressure.) Assuming that the entire fuel value of the hamburger is used to vaporize water, it will require that

$$\left(\frac{6.84 \times 10^3 \text{ kJ}}{44.01 \text{ kJ mol}^{-1}} \right) \left(\frac{18.02 \text{ g } H_2O}{1 \text{ mol } H_2O} \right) = 2.8 \times 10^3 \text{g } H_2O$$

be vaporized.

3.68   An oxyacetylene flame is often used in the welding of metals. Estimate the flame temperature produced by the reaction

$$2C_2H_2(g) + 5O_2(g) \rightarrow 4CO_2(g) + 2H_2O(g)$$

Assume the heat generated from this reaction is all used to heat the products. (*Hint:* First calculate the value of $\Delta_r H^\circ$ for the reaction. Next, look up the heat capacities of the products. Assume the heat capacities are temperature independent.)

The enthalpy of reaction for the oxidation of acetylene is

$$\Delta_r H^\circ = 4\Delta_f \overline{H}^\circ \left[CO_2(g) \right] + 2\Delta_f \overline{H}^\circ \left[H_2O(g) \right] - 2\Delta_f \overline{H}^\circ \left[C_2H_2(g) \right] - 5\Delta_f \overline{H}^\circ \left[O_2(g) \right]$$

$$= 4 \left(-393.5 \text{ kJ mol}^{-1} \right) + 2 \left(-241.8 \text{ kJ mol}^{-1} \right) - 2 \left(226.6 \text{ kJ mol}^{-1} \right) - 5 \left(0 \text{ kJ mol}^{-1} \right)$$

$$= -2510.8 \text{ kJ mol}^{-1}$$

Because the reaction takes place at constant pressure, the enthalpy of reaction is the same as heat released by the reaction, and this heat is absorbed by the products. The initial temperature, T_i, is assumed to be 298 K, and the final temperature of the products, T_f is

$$q = 2510.8 \text{ kJ mol}^{-1} = \left\{ 4\overline{C}_P^{\circ} \left[CO_2(g) \right] + 2\overline{C}_P^{\circ} \left[H_2O(g) \right] \right\} \left(T_f - T_i \right)$$

$$= \left[4 \left(37.1 \text{ J K}^{-1} \text{mol}^{-1} \right) + 2 \left(33.6 \text{ J K}^{-1} \text{mol}^{-1} \right) \right] \left(T_f - 298 \text{ K} \right) \left(\frac{1 \text{ kJ}}{1000 \text{ J}} \right)$$

$$T_f - 298 \text{ K} = 1.165 \times 10^4 \text{ K}$$

$$T_f = 1.19 \times 10^4 \text{ K}$$

The value of $\overline{C}_P^{\circ}$ for $H_2O(g)$ may be found in a standard reference.

3.70 The enthalpies of hydrogenation of ethylene and benzene have been determined at 298 K:

$$C_2H_4(g) + H_2(g) \rightarrow C_2H_6(g) \quad \Delta_r H^{\circ} = -132 \text{ kJ mol}^{-1}$$

$$C_6H_6(g) + 3H_2(g) \rightarrow C_6H_{12}(g) \quad \Delta_r H^{\circ} = -246 \text{ kJ mol}^{-1}$$

What would be the enthalpy of hydrogenation for benzene if it contained three isolated, unconjugated double bonds? How would you account for the difference between the calculated value based on this assumption and the measured value?

If benzene contained three isolated, unconjugated double bonds, its enthalpy of hydrogenation could be estimated as three times that for the single double bond in ethylene, or $\Delta_r H^{\circ}_{calc} = 3 \left(-132 \text{ kJ mol}^{-1} \right) = -396 \text{ kJ mol}^{-1}$. The difference is $-246 \text{ kJ mol}^{-1} - \left(-396 \text{ kJ mol}^{-1} \right) = 150 \text{ kJ mol}^{-1}$. This implies that benzene is 150 kJ mol^{-1} more stable than the hypothetical molecule with three isolated, unconjugated double bonds. This is attributed to the resonance (electron delocalization) energy of benzene.

3.72 The standard enthalpy of formation at 298 K of HF(aq) is $-320.1 \text{ kJ mol}^{-1}$; OH$^-$($aq$), -229.6 kJ mol^{-1}; F$^-$(aq), $-329.11 \text{ kJ mol}^{-1}$; and $H_2O(l)$, $-285.84 \text{ kJ mol}^{-1}$. **(a)** Calculate the enthalpy of neutralization of HF(aq),

$$HF(aq) + OH^-(aq) \rightarrow F^-(aq) + H_2O(l)$$

(b) Using the value of $-55.83 \text{ kJ mol}^{-1}$ as the enthalpy change from the reaction

$$H^+(aq) + OH^-(aq) \rightarrow H_2O(l)$$

calculate the enthalpy change for the dissociation

$$HF(aq) \rightarrow H^+(aq) + F^-(aq)$$

(a)

$$\Delta_r H^\circ = \Delta_f \overline{H}^\circ \left[F^-(aq) \right] + \Delta_f \overline{H}^\circ \left[H_2O(l) \right] - \Delta_f \overline{H}^\circ \left[HF(aq) \right] - \Delta_f \overline{H}^\circ \left[OH^-(aq) \right]$$

$$= -329.11 \text{ kJ mol}^{-1} + \left(-285.8 \text{ kJ mol}^{-1} \right) - \left(-320.1 \text{ kJ mol}^{-1} \right) - \left(-229.6 \text{ kJ mol}^{-1} \right)$$

$$= -65.2 \text{ kJ mol}^{-1}$$

(b) The dissociation of HF can be considered as a sum of the following equations:

$$HF(aq) + OH^-(aq) \rightarrow F^-(aq) + H_2O(l) \qquad \Delta_r H^\circ = -65.2 \text{ kJ mol}^{-1}$$

$$H_2O(l) \rightarrow H^+(aq) + OH^-(aq) \qquad \Delta_r H^\circ = 55.83 \text{ kJ mol}^{-1}$$

The enthalpy change for the dissociation reaction is therefore

$$\Delta_r H^\circ = -65.2 \text{ kJ mol}^{-1} + 55.83 \text{ kJ mol}^{-1} = -9.4 \text{ kJ mol}^{-1}$$

3.74 Metabolic activity in the human body releases approximately 1.0×10^4 kJ of heat per day. Assuming the body is 50 kg of water, how fast would the body temperature rise if it were an isolated system? How much water must the body eliminate as perspiration to maintain the normal body temperature (98.6° F)? Comment on your results. The heat of vaporization of water may be taken as 2.41 kJ g^{-1}.

If the body absorbs all the heat released and is an isolated system, then the temperature rise, ΔT, is related to q in the following fashion:

$$q_{absorbed} = C_P \left[H_2O(l) \right] \Delta T$$

$$\Delta T = \frac{q}{C_P \left[H_2O(l) \right]} = \frac{\left(1.0 \times 10^4 \text{ kJ}\right) \left(\frac{1000 \text{ J}}{1 \text{ kJ}} \right)}{\left(\frac{50 \times 10^3 \text{ g}}{18.01 \text{ g mol}^{-1}} \right) \left(75.3 \text{ J K}^{-1} \text{mol}^{-1} \right)} = 47.8 \text{ K}$$

If the body temperature is to remain constant, then the heat released by metabolic activity must be used for the evaporation of water as perspiration, that is,

$$1.0 \times 10^4 \text{ kJ} = m_{water} \left(2.41 \text{ kJ g}^{-1} \right)$$

$$m_{water} = 4.1 \times 10^3 \text{ g}$$

The actual amount of perspiration is less than this because part of the body heat is lost to the surroundings by convection and radiation.

3.76 Calculate the fraction of the enthalpy of vaporization of water used for the expansion of steam at its normal boiling point.

Treating the steam as an ideal gas, the molar volume of steam at 373 K is calculated as

$$\overline{V} = \frac{RT}{P}$$

$$= \frac{\left(0.08206 \text{ L atm K}^{-1} \text{ mol}^{-1}\right)(373 \text{ K})}{1 \text{ atm}}$$

$$= 30.61 \text{ L mol}^{-1}$$

The work done in the expansion from liquid water to this volume of steam is then found having made the assumption that the volume of the condensed phase is negligible. (One mole of liquid water has a volume of 18 mL, so the approximation is a good one.)

$$w = -P_{ex}\Delta \overline{V}$$

$$= -(1 \text{ atm})\left(30.61 \text{ L mol}^{-1}\right)\left(\frac{101.3 \text{ J}}{1 \text{ L atm}}\right)\left(\frac{1 \text{ kJ}}{1000 \text{ J}}\right)$$

$$= -3.101 \text{ kJ mol}^{-1}$$

At 373 K the molar enthalpy of vaporization of water is $\Delta_{vap}\overline{H}^{\circ} = 40.79 \text{ kJ mol}^{-1}$, so the fraction used for the expansion of steam is

$$\frac{3.101 \text{ kJ mol}^{-1}}{40.79 \text{ kJ mol}^{-1}} = 7.60\%$$

3.78 Calculate the internal energy of a Goodyear blimp filled with helium gas at 1.2×10^5 Pa (compared to the empty blimp). The volume of the inflated blimp is $5.5 \times 10^3 \text{ m}^3$. If all the energy were used to heat 10.0 tons of copper at $21°$ C, calculate the final temperature of the metal. (*Hint*: 1 ton $= 9.072 \times 10^5$ g.)

The internal energy of a monoatomic (ideal) gas is $\frac{3}{2}nRT$. Likewise for an ideal gas, $PV = nRT$. Thus, for an ideal gas, $U = \frac{3}{2}PV$. For the blimp,

$$U = \frac{3}{2}\left(1.2 \times 10^5 \text{ Pa}\right)\left(5.5 \times 10^3 \text{ m}^3\right) = 9.90 \times 10^8 \text{ J}$$

If all this energy is used to heat copper, the temperature change is related to the amount of energy used via $q = m_{Cu}s_{Cu}\Delta T$, giving

$$9.90 \times 10^8 \text{ J} = (10.0 \text{ ton})\left(\frac{9.072 \times 10^5 \text{ g}}{1 \text{ ton}}\right)\left(24.47 \text{ J} °\text{C}^{-1} \text{ mol}^{-1}\right)\left(\frac{1 \text{ mol Cu}}{63.55 \text{ g Cu}}\right)\Delta T$$

$$9.90 \times 10^8 = 3.49 \times 10^6 °\text{C}^{-1}\Delta T$$

$$283.7° \text{ C} = \Delta T = T_f - 21° \text{ C}$$

Thus, $T_f = 305°$ C

3.80 An ideal gas is isothermally compressed from P_1, V_1 to P_2, V_2. Under what conditions would the work done be a minimum? A maximum? Write the expressions for minimum and maximum work done for this process. Explain your reasoning.

The conditions are opposite to those for the isothermal *expansion* of an ideal gas. The minimum work is done in an reversible process because the opposing (internal) pressure is only infinitesimally smaller than the external pressure causing the compression. The work done under these conditions is

$$w_{min} = -nRT \ln \frac{V_2}{V_1}$$

Since $V_2 < V_1$, the work is positive as expected for a compression.

The maximum work would be done in an irreversible process and an external pressure of P_2, giving a value of

$$w_{max} = -P_2(V_2 - V_1)$$

3.82   State whether each of the following statements is true or false: **(a)** $\Delta U \approx \Delta H$ except for gases or high-pressure processes. **(b)** In gas compression, a reversible process does maximum work. **(c)** ΔU is a state function. **(d)** $\Delta U = q + w$ for an open system. **(e)** C_V is temperature independent for gases. **(f)** The internal energy of a real gas depends only on temperature.

(a) True **(b)** False **(c)** False (U is a state function, ΔU is not.) **(d)** False (It is true for a *closed* system.) **(e)** False **(f)** False

---

3.84   Derive an expression for the work done during the isothermal, reversible expansion of a van der Waals gas. Account physically for the way in which the coefficients a and b appear in the final expression. [*Hint*: You need to apply the Taylor series expansion:

$$\ln(1 - x) = -x - \frac{x^2}{2} \cdots \qquad \text{for } |x| \ll 1$$

to the expression $\ln(V - nb)$. Recall that the a term represents attraction and the b term repulsion.]

$$w = -\int_{V_1}^{V_2} P \, dV$$

$$= -\int_{V_1}^{V_2} \left(\frac{nRT}{V - nb} - \frac{an^2}{V^2} \right) dV$$

$$= -nRT \ln \frac{V_2 - nb}{V_1 - nb} - an^2 \left(\frac{1}{V_2} - \frac{1}{V_1} \right)$$

The ln term can be written as

$$\ln \frac{V_2 - nb}{V_1 - nb} = \ln \frac{V_2 \left(1 - \frac{nb}{V_2}\right)}{V_1 \left(1 - \frac{nb}{V_1}\right)}$$

$$= \ln \frac{V_2}{V_1} + \ln \frac{1 - \frac{nb}{V_2}}{1 - \frac{nb}{V_1}}$$

$$= \ln \frac{V_2}{V_1} + \ln \left(1 - \frac{nb}{V_2}\right) - \ln \left(1 - \frac{nb}{V_1}\right)$$

Assume the volume occupied by the gas molecules is much greater than the volume of the molecules, both before and after the expansion, that is, $V_1 \gg nb$ and $V_2 \gg nb$, then $\frac{nb}{V_1} \ll 1$ and $\frac{nb}{V_2} \ll 1$. Under these condition, the Taylor expansion described in the question can be used to simplify the ln terms.

$$\ln \frac{V_2 - nb}{V_1 - nb} = \ln \frac{V_2}{V_1} + \left(-\frac{nb}{V_2} - \frac{1}{2}\frac{n^2 b^2}{V_2^2} + \cdots\right) - \left(-\frac{nb}{V_1} - \frac{1}{2}\frac{n^2 b^2}{V_1^2} + \cdots\right)$$

$$= \ln \frac{V_2}{V_1} - nb \left(\frac{1}{V_2} - \frac{1}{V_1}\right) - \frac{n^2 b^2}{2}\left(\frac{1}{V_2^2} - \frac{1}{V_1^2}\right) + \cdots$$

Substitute the above expression into w.

$$w = -nRT \left[\ln \frac{V_2}{V_1} - nb \left(\frac{1}{V_2} - \frac{1}{V_1}\right) - \frac{n^2 b^2}{2}\left(\frac{1}{V_2^2} - \frac{1}{V_1^2}\right) + \cdots\right] - an^2 \left(\frac{1}{V_2} - \frac{1}{V_1}\right)$$

$$= -nRT \ln \frac{V_2}{V_1} + (bRT - a) n^2 \left(\frac{1}{V_2} - \frac{1}{V_1}\right) + \frac{n^3 b^2 RT}{2}\left(\frac{1}{V_2^2} - \frac{1}{V_1^2}\right) + \cdots$$

The first term in this last equation is just the result for the ideal gas. It is modified by the succeeding terms to account for intermolecular interactions. The second term shows the balance between attractive and repulsive forces. In an expansion, $V_2 > V_1$ so $\left(\frac{1}{V_2} - \frac{1}{V_1}\right) < 0$. If attractive forces dominate, $a > bRT$ and the entire second term is positive, which cancels some of the (negative) work done in the expansion. Because of the attractive forces, some energy must be used to overcome the intermolecular interactions, and not as much work can be done as in the case of the ideal gas.

On the other hand, if $a < bRT$, the entire second term is negative and enhances the (negative) work done in the expansion. In this case the repulsive forces dominate, and the energy released as the molecules move farther apart from each other is available to do more work than the ideal gas could. The higher order terms reinforce this effect for high densities where the repulsive forces are most significant.

3.86 One mole of ammonia initially at $5°$ C is placed in contact with 3 moles of helium initially at $90°$ C. Given that $\overline{C}_V$ for ammonia is $3R$, if the process is carried out at constant total volume, what is the final temperature of the gases?

To reach equilibrium, ammonia will warm up and helium will cool down until both reach the same final temperature, T_f. According to the law of conservation of energy, the energy given up by helium must be absorbed by ammonia, assuming energy is not transferred to the surroundings. This energy can be calculated using heat capacity. Heat capacity is the energy required to raise the temperature of a given quantity of substance by $1°$ C or 1 K. Therefore, the energy required to warm up ammonia is

$$n_{NH_3} \overline{C}_{V, NH_3} \left(T_f - 5° \text{ C} \right)$$

and the energy removed in cooling He is

$$-n_{He} \overline{C}_{V, He} \left(T_f - 90° \text{ C} \right)$$

(The minus sign indicates that energy is being lost.)

Conservation of energy requires that these energies be equal. Using $\overline{C}_{V, He} = \frac{3}{2} R$ (He has only translational degrees of freedom, and therefore, $U = \frac{3}{2} RT$.) and the given value for $\overline{C}_{V, NH_3}$,

$$n_{NH_3} \overline{C}_{V, NH_3} \left(T_f - 5° \text{ C} \right) = -n_{He} \overline{C}_{V, He} \left(T_f - 90° \text{ C} \right)$$

$$(1 \text{ mol}) (3R) \left(T_f - 5° \text{ C} \right) = (3 \text{ mol}) \left(\frac{3}{2} R \right) \left(90° \text{ C} - T_f \right)$$

$$\left(T_f - 5° \text{ C} \right) = \frac{3}{2} \left(90° \text{ C} - T_f \right)$$

$$T_f - 5° \text{ C} = 135° \text{ C} - \frac{3}{2} T_f$$

$$\frac{5}{2} T_f = 140° \text{ C}$$

$$T_f = 56° \text{ C}$$

3.88   The first excited electronic energy level of the helium atom is 3.13×10^{-18} J above the ground level. Estimate the temperature at which the electronic motion will begin to make a significant contribution to the heat capacity. That is, at what temperature will the ratio of the population of the first excited state to the ground state be 5.0%?

Find the temperature, T, such that the population ratio, $N_2/N_1 = 0.050$,

$$\frac{N_2}{N_1} = e^{-\frac{\Delta E}{k_B T}} = 0.050$$

$$-\frac{\Delta E}{k_B T} = \ln 0.050$$

$$T = -\frac{\Delta E}{k_B \ln 0.050} = -\frac{3.13 \times 10^{-18} \text{ J}}{\left(1.381 \times 10^{-23} \text{ J K}^{-1} \right) (\ln 0.050)} = 7.6 \times 10^4 \text{ K}$$

3.90   From your knowledge of heat capacity, explain why hot, humid air is more uncomfortable than hot, dry air and cold, damp air is more uncomfortable than cold, dry air.

As a non-linear, triatomic molecule, H_2O has more rotational motions (3 versus 2) than the two diatomic molecules, N_2 and O_2, that comprise the bulk of air. This leads to a larger heat capacity for water compared to the other molecules, which means that more energy is required to effect a given temperature change. Thus, hot, humid air with its greater content of water vapor, is capable of transferring more thermal energy to your body than the diatomic gases, making hot, humid air more uncomfortable than hot, dry air. (Of course, the retardation of the evaporation rate from your skin is another significant reason.) On the other hand, the greater water content in cold damp air is more able to remove thermal energy from your body than cold, dry air can.

3.92   Give an interpretation for the following DSC thermogram for the thermal denaturation of a protein.

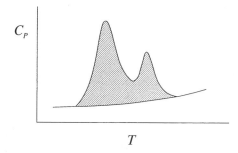

The two maxima in the C_P versus T plot indicate that the protein denaturation occurs in two steps.

The Second Law of Thermodynamics

PROBLEMS AND SOLUTIONS

4.2 One of the many statements of the second law of thermodynamics is: Heat cannot flow from a colder body to a warmer one without external aid. Assume two systems, 1 and 2, at T_1 and T_2 ($T_2 > T_1$). Show that if a quantity of heat q did flow spontaneously from 1 to 2, the process would result in a decrease in entropy of the universe. (You may assume that the heat flows very slowly so that the process can be regarded as reversible. Assume also that the loss of heat by system 1 and the gain of heat by system 2 do not affect T_1 and T_2.)

The universe is comprised of the two systems. Since heat flows from system 1 to 2, $q_1 = -q_{rev}$ and $q_2 = q_{rev}$.

$$\Delta S_{univ} = \Delta S_1 + \Delta S_2 = \frac{q_1}{T_1} + \frac{q_2}{T_2} = \frac{-q_{rev}}{T_1} + \frac{q_{rev}}{T_2}$$

$$= q_{rev}\left(\frac{1}{T_2} - \frac{1}{T_1}\right) < 0 \qquad \text{since } T_1 < T_2$$

4.4 Molecules of a gas at any temperature T above the absolute zero are in constant motion. Does this "perpetual motion" violate the laws of thermodynamics?

As long as the gas does no work on the surroundings it violates no law of thermodynamics. Were the gas to do work on the surroundings, then the laws of thermodynamics would require that $w = \Delta U - q$ (the first law) and $\Delta S \geq \frac{q}{T}$ for an isothermal process (the second law). This last requirement means that the state of the system must change when work is done, since for an isothermal process doing work on the surroundings, $\Delta U = 0$ (assuming an ideal gas) and $w < 0$ imply $q > 0$. Thus, $\Delta S > 0$ and the system must have changed state, since S is a state function.

4.6 On a hot summer day, a person tries to cool himself by opening the door of a refrigerator. Is this a wise action, thermodynamically speaking?

No, this is not wise. The refrigerator is not working under ideal conditions, so that it delivers more heat to the room than it extracts. Indeed, even under ideal conditions, the second law of thermodynamics requires an excess of heat delivered to the room over that extracted. The room will get warmer than if the door remained closed.

4.8   Calculate the values of ΔU, ΔH, and ΔS for the following process:

1 mole of liquid water at 25° C and 1 atm → 1 mol of steam at 100° C and 1 atm

The molar heat of vaporization of water at 373 K is 40.79 kJ mol^{-1}, and the molar heat capacity of water is 75.3 J K^{-1} mol^{-1}. Assume the molar heat capacity to be temperature independent and ideal-gas behavior.

The problem can be solved readily by breaking down the process into two steps, each carried out at 1 atm: (1) $H_2O(l)$ is heated from 25° C to 100° C, then (2) $H_2O(l)$ at 100° C is heated to effect the phase transformation to $H_2O(g)$ at 100° C.

Step 1

$$H_2O(l),\ 25°\,C \rightarrow H_2O(l),\ 100°\,C$$

$$\Delta H = C_P \Delta T = (1\ \text{mol}) \left(75.3\ \text{J K}^{-1}\text{mol}^{-1}\right)(373.2\ \text{K} - 298.2\ \text{K}) = 5.648 \times 10^3\ \text{J}$$

ΔU is related to ΔH by

$$\Delta U = \Delta H - P\Delta V$$

Since both reactant and product are in the liquid phase, ΔV is negligible. Therefore,

$$\Delta U = \Delta H = 5.648 \times 10^3\ \text{J}$$

$$\Delta S = C_P \ln \frac{T_2}{T_1} = (1\ \text{mol})\left(75.3\ \text{J K}^{-1}\ \text{mol}^{-1}\right) \ln \frac{373.2\ \text{K}}{298.2\ \text{K}} = 16.89\ \text{J K}^{-1}$$

Step 2

$$H_2O(l),\ 100°\,C \rightarrow H_2O(g),\ 100°\,C$$

$$\Delta H = n\Delta_{\text{vap}}\overline{H} = (1\ \text{mol})\left(40.79\ \text{kJ mol}^{-1}\right) = 40.79\ \text{kJ}$$

To calculate ΔU, the change in volume must first be determined. Since the volume of $H_2O(g)$, V_g is much greater than that of $H_2O(l)$, V_l, the latter is ignored.

$$\Delta U = \Delta H - P\Delta V = \Delta H - P\left(V_g - V_l\right) = \Delta H - PV_g$$

$$= \Delta H - P\frac{nRT}{P} = \Delta H - nRT$$

$$= 40.79\ \text{kJ} - (1\ \text{mol})\left(8.314\ \text{J mol}^{-1}\ \text{K}^{-1}\right)(373.2\ \text{K})\left(\frac{1\ \text{kJ}}{1000\ \text{J}}\right)$$

$$= 37.687\ \text{kJ}$$

$$\Delta S = \frac{\Delta_{\text{vap}}H}{T_b} = \frac{40.79 \times 10^3\ \text{J}}{373.15\ \text{K}} = 109.31\ \text{J K}^{-1}$$

The values of ΔH, ΔU, and ΔS for the entire process can be obtained by summing the corresponding quantities calculated in the two steps:

$$\Delta H = 5.648 \text{ kJ} + 40.79 \text{ kJ} = 46.44 \text{ kJ}$$

$$\Delta U = 5.648 \text{ kJ} + 37.687 \text{ kJ} = 43.34 \text{ kJ}$$

$$\Delta S = 16.89 \text{ J K}^{-1} + 109.31 \text{ J K}^{-1} = 126.2 \text{ J K}^{-1}$$

4.10 A quantity of 6.0 moles of an ideal gas is reversibly heated at constant volume from $17°$ C to $35°$ C. Calculate the entropy change. What would be the value of ΔS if the heating were carried out irreversibly?

At constant volume, $dq_{\text{rev}} = C_V dT$.

$$\Delta S = \int \frac{dq_{\text{rev}}}{T} = \int \frac{C_V dT}{T} = C_V \ln \frac{T_2}{T_1} = \frac{3}{2} nR \ln \frac{T_2}{T_1}$$

$$= \frac{3}{2} (6.0 \text{ mol}) \left(8.314 \text{ J K}^{-1} \text{mol}^{-1} \right) \ln \frac{308 \text{ K}}{290 \text{ K}}$$

$$= 4.5 \text{ J K}^{-1}$$

If heating were carried out irreversibly, ΔS is still 4.5 J K^{-1} because S is a state function so that ΔS depends only on the final and initial states. ΔS must be calculated, however, using a reversible pathway.

4.12 A quantity of 35.0 g of water at $25.0°$ C (called A) is mixed with 160.0 g of water at $86.0°$ C (called B). **(a)** Calculate the final temperature of the system, assuming that the mixing is carried out adiabatically. **(b)** Calculate the entropy change of A, B, and the entire system.

(a) Let the final temperature be T_f and the specific heat of water be s. Since the process is carried out adiabatically, the energy entering A as heat, q_A, is equal in magnitude but opposite in sign to that leaving B as heat, q_B.

$$q_A = -q_B$$

$$(35.0 \text{ g}) \, s \, \left(T_f - 25.0° \text{ C} \right) = - \left(160.0 \text{ g} \right) s \, \left(T_f - 86.0° \text{ C} \right)$$

$$T_f - 25.0° \text{ C} = -4.571 \left(T_f - 86.0° \text{ C} \right)$$

$$= -4.571 T_f + 393.1° \text{ C}$$

$$5.571 T_f = 418.1° \text{ C}$$

$$T_f = 75.0° \text{ C}$$

(b) Let the entropy change of A be ΔS_A and the entropy change of B be ΔS_B.

$$\Delta S_A = C_P \ln \frac{348.2 \text{ K}}{298.2 \text{ K}} = \left(\frac{35.0 \text{ g}}{18.02 \text{ g mol}^{-1}} \right) \left(75.3 \text{ J K}^{-1} \text{mol}^{-1} \right) \ln \frac{348.2 \text{ K}}{298.2 \text{ K}} = 22.7 \text{ J K}^{-1}$$

$$\Delta S_B = C_P \ln \frac{348.2 \text{ K}}{359.2 \text{ K}} = \left(\frac{160.0 \text{ g}}{18.02 \text{ g mol}^{-1}} \right) \left(75.3 \text{ J K}^{-1} \text{mol}^{-1} \right) \ln \frac{348.2 \text{ K}}{359.2 \text{ K}} = -20.79 \text{ J K}^{-1}$$

$$\Delta S_{\text{total}} = \Delta S_A + \Delta S_B = 22.7 \text{ J K}^{-1} - 20.79 \text{ J K}^{-1} = 1.9 \text{ J K}^{-1}$$

4.14 A sample of neon (Ne) gas initially at 20° C and 1.0 atm is expanded from 1.2 L to 2.6 L and simultaneously heated to 40° C. Calculate the entropy change for the process.

The number of moles of Ne can be determined using the initial conditions and the ideal gas law.

$$n = \frac{(1.0 \text{ atm})(1.2 \text{ L})}{(0.08206 \text{ L atm K}^{-1} \text{ mol}^{-1})(293 \text{ K})} = 4.99 \times 10^{-2} \text{ mol}$$

The problem can be solved by breaking down the process into 2 steps: (1) isothermal expansion from 1.2 L at 1.0 atm to 2.6 L. The temperature is kept at 20° C; (2) heating at constant volume (2.6 L) from 20° C to 40° C. The entropy changes for these two steps, ΔS_1 and ΔS_2 are

$$\Delta S_1 = nR \ln \frac{V_2}{V_1} = \left(4.99 \times 10^{-2} \text{ mol}\right)\left(8.314 \text{ J K}^{-1} \text{ mol}^{-1}\right) \ln \frac{2.6 \text{ L}}{1.2 \text{ L}} = 0.321 \text{ J K}^{-1}$$

$$\Delta S_2 = C_V \ln \frac{T_2}{T_1} = \frac{3}{2} nR \ln \frac{T_2}{T_1} = \frac{3}{2}\left(4.99 \times 10^{-2} \text{ mol}\right)\left(8.314 \text{ J K}^{-1} \text{ mol}^{-1}\right) \ln \frac{313 \text{ K}}{293 \text{ K}}$$

$$= 4.11 \times 10^{-2} \text{ J K}^{-1}$$

The entropy change for the entire process is

$$\Delta S = \Delta S_1 + \Delta S_2 = 0.321 \text{ J K}^{-1} + 4.11 \times 10^{-2} \text{ J K}^{-1} = 0.36 \text{ J K}^{-1}$$

4.16 One mole of an ideal gas at 298 K expands isothermally from 1.0 L to 2.0 L **(a)** reversibly and **(b)** against a constant external pressure of 12.2 atm. Calculate the values of ΔS_{sys}, ΔS_{surr}, and ΔS_{univ} in both cases. Are your results consist with the nature of the processes?

(a)

$$\Delta S_{sys} = nR \ln \frac{V_2}{V_1} = (1 \text{ mol})\left(8.314 \text{ J K}^{-1} \text{ mol}^{-1}\right) \ln \frac{2.0 \text{ L}}{1.0 \text{ L}} = 5.8 \text{ J K}^{-1}$$

$$\Delta S_{surr} = -5.8 \text{ J K}^{-1}$$

$$\Delta S_{univ} = 0 \text{ J K}^{-1}$$

(b) ΔS_{sys} is the same above, that is, 5.8 J K^{-1}, since S is a state function, although ΔS has to be calculated using a reversible path.

ΔS_{surr} can be calculated once q_{surr} is determined. The latter quantity is related to q_{sys}, which in turn can be calculated from the work done by the system, w, and the first law of thermodynamics.

$$w = -P_{ex}\Delta V = -(12.2 \text{ atm})(2.0 \text{ L} - 1.0 \text{ L})\left(\frac{101.3 \text{ J}}{1 \text{ L atm}}\right) = -1.236 \times 10^3 \text{ J}$$

According to the first law, $\Delta U = q + w$. Since $\Delta U = 0$ for an isothermal process, $q_{sys} = q = -w = 1.236 \times 10^3$ J. The entropy change of the surroundings is

$$\Delta S_{surr} = \frac{q_{surr}}{T_{surr}} = \frac{-q_{sys}}{T_{surr}} = \frac{-1.236 \times 10^3 \text{ J}}{298 \text{ K}} = -4.15 \text{ J K}^{-1}$$

Now ΔS_{univ} can be determined:

$$\Delta S_{univ} = 5.8 \text{ J K}^{-1} - 4.15 \text{ J K}^{-1} = 1.7 \text{ J K}^{-1}$$

The results in both parts are consistent with the nature of the processes. Specifically, for a reversible process, $\Delta S_{univ} = 0$, whereas for a spontaneous process, $\Delta S_{univ} > 0$.

4.18 A quantity of 0.54 mole of steam initially at 350° C and 2.4 atm undergoes a cyclic process for which $q = -74$ J. Calculate the value of ΔS for the process.

$\Delta S = 0$ for a cyclic process.

4.20 Use the data in Appendix 2 to calculate the values of $\Delta_r S^o$ of the reactions listed in the previous problem.

(a)

$$\Delta_r S^o = 2\overline{S}^o \left[Fe_2O_3(s) \right] - 4\overline{S}^o \left[Fe(s) \right] - 3\overline{S}^o \left[O_2(g) \right]$$

$$= 2 \left(90.0 \text{ J K}^{-1} \text{mol}^{-1} \right) - 4 \left(27.2 \text{ J K}^{-1} \text{mol}^{-1} \right) - 3 \left(205.0 \text{ J K}^{-1} \text{mol}^{-1} \right)$$

$$= -543.8 \text{ J K}^{-1} \text{mol}^{-1}$$

(b)

$$\Delta_r S^o = \overline{S}^o \left[O_2(g) \right] - 2\overline{S}^o \left[O(g) \right]$$

$$= 205.0 \text{ J K}^{-1} \text{mol}^{-1} - 2 \left(161.0 \text{ J K}^{-1} \text{mol}^{-1} \right)$$

$$= -117.0 \text{ J K}^{-1} \text{mol}^{-1}$$

(c)

$$\Delta_r S^o = \overline{S}^o \left[NH_3(g) \right] + \overline{S}^o \left[HCl(g) \right] - \overline{S}^o \left[NH_4Cl(s) \right]$$

$$= 192.5 \text{ J K}^{-1} \text{mol}^{-1} + 186.5 \text{ J K}^{-1} \text{mol}^{-1} - 94.6 \text{ J K}^{-1} \text{mol}^{-1}$$

$$= 284.4 \text{ J K}^{-1} \text{mol}^{-1}$$

(d)

$$\Delta_r S^o = 2\overline{S}^o \left[HCl(g) \right] - \overline{S}^o \left[H_2(g) \right] - \overline{S}^o \left[Cl_2(g) \right]$$

$$= 2 \left(186.5 \text{ J K}^{-1} \text{mol}^{-1} \right) - \left(130.6 \text{ J K}^{-1} \text{mol}^{-1} \right) - \left(223.0 \text{ J K}^{-1} \text{mol}^{-1} \right)$$

$$= 19.4 \text{ J K}^{-1} \text{mol}^{-1}$$

These results agree with predictions made in the previous problem.

4.22 One mole of an ideal gas is isothermally expanded from 5.0 L to 10 L at 300 K. Compare the entropy changes for the system, surroundings, and the universe if the process is carried out **(a)** reversibly, and **(b)** irreversibly against an external pressure of 2.0 atm.

(a) For the reversible process,

$$\Delta S_{sys} = nR \ln \frac{V_2}{V_1} = (1 \text{ mol}) \left(8.314 \text{ J K}^{-1} \text{mol}^{-1} \right) \ln \frac{10 \text{ L}}{5.0 \text{ L}} = 5.8 \text{ J K}^{-1}$$

$$\Delta S_{surr} = -5.8 \text{ J K}^{-1}$$

$$\Delta S_{univ} = 0 \text{ J K}^{-1}$$

(b) ΔS_{sys} is the same as above, that is, 5.8 J K^{-1}.

ΔS_{surr} can be calculated once q_{surr} is determined. The latter quantity is related to q_{sys}, which in turn can be calculated from the work done by the system, w, and the first law of thermodynamics.

$$w = -P_{ex}\Delta V = -(2.0 \text{ atm})(10 \text{ L} - 5.0 \text{ L})\left(\frac{101.3 \text{ J}}{1 \text{ L atm}}\right) = -1.01 \times 10^3 \text{ J}$$

According to the first law, $\Delta U = q + w$. For an ideal gas, $\Delta U = 0$ for an isothermal process, and $q_{sys} = q = -w = 1.01 \times 10^3$ J. The entropy change of the surroundings is

$$\Delta S_{surr} = \frac{q_{surr}}{T_{surr}} = \frac{-q_{sys}}{T_{surr}} = \frac{-1.01 \times 10^3 \text{ J}}{300 \text{ K}} = -3.4 \text{ J K}^{-1}$$

Therefore,

$$\Delta S_{univ} = 5.8 \text{ J K}^{-1} - 3.4 \text{ J K}^{-1} = 2.4 \text{ J K}^{-1}$$

4.24   Consider the reaction

$$N_2(g) + O_2(g) \rightarrow 2NO(g)$$

Calculate the values of $\Delta_r S^\circ$ for the reaction mixture, surroundings, and the universe at 298 K. Why is your result reassuring to Earth's inhabitants?

$$\Delta_r S^\circ = 2\overline{S}^\circ\left[NO(g)\right] - \overline{S}^\circ\left[N_2(g)\right] - \overline{S}^\circ\left[O_2(g)\right]$$

$$= 2\left(210.6 \text{ J K}^{-1}\text{ mol}^{-1}\right) - 191.6 \text{ J K}^{-1}\text{ mol}^{-1} - 205.0 \text{ J K}^{-1}\text{ mol}^{-1}$$

$$= 24.6 \text{ J K}^{-1}\text{ mol}^{-1}$$

ΔS°_{surr} is determined from $\Delta_r H^\circ$ and the temperature of the surroundings.

$$\Delta_r H^\circ = 2\Delta_f\overline{H}^\circ\left[NO(g)\right] - \Delta_f\overline{H}^\circ\left[N_2(g)\right] - \Delta_f\overline{H}^\circ\left[O_2(g)\right]$$

$$= 2\left(90.4 \text{ kJ mol}^{-1}\right) - 0 \text{ kJ mol}^{-1} - 0 \text{ kJ mol}^{-1}$$

$$= 180.8 \text{ kJ mol}^{-1}$$

$$\Delta H^\circ_{surr} = -\Delta_r H^\circ = -180.8 \text{ kJ mol}^{-1}$$

$$\Delta S^\circ_{surr} = \frac{\Delta H^\circ_{surr}}{T} = \frac{-180.8 \times 10^3 \text{ J mol}^{-1}}{298 \text{ K}} = -607 \text{ J K}^{-1}\text{ mol}^{-1}$$

Therefore,

$$\Delta S_{univ} = 24.6 \text{ J K}^{-1}\text{ mol}^{-1} - 607 \text{ J K}^{-1}\text{ mol}^{-1} = -582 \text{ J K}^{-1}\text{ mol}^{-1}$$

This is not a spontaneous process at 298 K. Therefore, O_2, which is essential to us, does not react with N_2 in the atmosphere at 298 K.

4.26 Choose the substance with the greater molar entropy in each of the following pairs: **(a)** $H_2O(l)$, $H_2O(g)$, **(b)** $NaCl(s)$, $CaCl_2(s)$, **(c)** $N_2(0.1 \text{ atm})$, $N_2(1 \text{ atm})$, **(d)** C(diamond), C(graphite), **(e)** $O_2(g)$, $O_3(g)$, **(f)** ethanol (C_2H_5OH), dimethyl ether (C_2H_6O), **(g)** $N_2O_4(g)$, $2NO_2(g)$, and **(h)** Fe(s) at 298 K, Fe(s) at 398 K. (Unless otherwise stated, assume the temperature is 298 K.)

(a) $H_2O(g)$, a gas has greater entropy than the more ordered liquid.

(b) $CaCl_2(s)$, this is a more complex system than NaCl(s).

(c) N_2 (0.1 atm), at the lower pressure, the gas occupies a larger volume leading to a larger number of microstates for the system.

(d) C(graphite), diamond is a more ordered solid than is graphite.

(e) $O_3(g)$, this is a more complex system than diatomic $O_2(g)$.

(f) Dimethyl ether, ethanol can form hydrogen bonds leading to a more ordered system.

(g) $N_2O_4(g)$, one mole of $N_2O_4(g)$ is a more complex system and has greater entropy than one mole of $NO_2(g)$, although *two* moles of $NO_2(g)$ has greater entropy than one mole of $N_2O_4(g)$.

(h) Fe(s) at 398 K, since it is at a higher temperature.

4.28 A quantity of 0.35 mole of an ideal gas initially at 15.6° C is expanded from 1.2 L to 7.4 L. Calculate the values of w, q, ΔU, ΔS, and ΔG if the process is carried out **(a)** isothermally and reversibly, and **(b)** isothermally and irreversibly against an external pressure of 1.0 atm.

(a) For an isothermal process of an ideal gas, $\Delta U = 0$ and $\Delta H = 0$.

$$w = -nRT \ln \frac{V_2}{V_1} = -(0.35 \text{ mol}) \left(8.314 \text{ J K}^{-1} \text{mol}^{-1}\right) (288.8 \text{ K}) \ln \frac{7.4 \text{ L}}{1.2 \text{ L}}$$

$$= -1.53 \times 10^3 \text{ J} = -1.5 \times 10^3 \text{ J}$$

$$q = \Delta U - w = 1.53 \times 10^3 \text{ J} = 1.5 \times 10^3 \text{ J}$$

$$\Delta S = \frac{q_{rev}}{T} = \frac{1.53 \times 10^3 \text{ J}}{288.8 \text{ K}} = 5.30 \text{ J K}^{-1} = 5.3 \text{ J K}^{-1}$$

$$\Delta G = \Delta H - T\Delta S = 0 - (288.8 \text{ K}) \left(5.30 \text{ J K}^{-1}\right) = -1.5 \times 10^3 \text{ J}$$

(b) Since U, S, and G are state functions, ΔU, ΔS, ΔG depend only on the initial and final states and not on the path. Therefore, they are the same as above, that is

$$\Delta U = 0 \text{ J}$$

$$\Delta S = 5.3 \text{ J K}^{-1}$$

$$\Delta G = -1.5 \times 10^3 \text{ J}$$

w and q, however, are path dependent.

$$w = -P_{ex}\Delta V = -(1.0 \text{ atm}) (7.4 \text{ L} - 1.2 \text{ L}) \left(\frac{101.3 \text{ J}}{1 \text{ L atm}}\right) = -6.3 \times 10^2 \text{ J}$$

$$q = \Delta U - w = 6.3 \times 10^2 \text{ J}$$

4.30   Use the values listed in Appendix 2 to calculate the value of $\Delta_r G^\circ$ for the following alcohol fermentation:

$$\alpha\text{-D-glucose}(aq) \rightarrow 2C_2H_5OH(l) + 2CO_2(g)$$

$(\Delta_f \overline{G}^\circ[\alpha\text{-D-glucose}(aq)] = -914.5 \text{ kJ mol}^{-1})$

$$\Delta_r G^\circ = 2\Delta_f \overline{G}^\circ \left[C_2H_5OH(l) \right] + 2\Delta_f \overline{G}^\circ \left[CO_2(g) \right] - \Delta_f \overline{G}^\circ \left[\alpha\text{-D-glucose}(aq) \right]$$

$$= 2\left(-174.2 \text{ kJ mol}^{-1} \right) + 2 \left(-394.4 \text{ kJ mol}^{-1} \right) - \left(-914.5 \text{ kJ mol}^{-1} \right)$$

$$= -222.7 \text{ kJ mol}^{-1}$$

4.32   Certain bacteria in the soil obtain the necessary energy for growth by oxidizing nitrite to nitrate:

$$2NO_2^-(aq) + O_2(g) \rightarrow 2NO_3^-(aq)$$

Given that the standard Gibbs energies of formation of NO_2^- and NO_3^- are $-34.6 \text{ kJ mol}^{-1}$ and $-110.5 \text{ kJ mol}^{-1}$, respectively, calculate the amount of Gibbs energy released when 1 mole of NO_2^- is oxidized to 1 mole of NO_3^-.

According to the chemical equation $2NO_2^-(aq) + O_2(g) \rightarrow 2NO_3^-(aq)$,

$$\Delta_r G^\circ = 2\Delta_f \overline{G}^\circ \left[NO_3^-(aq) \right] - 2\Delta_f \overline{G}^\circ \left[NO_2^-(aq) \right] - \Delta_f \overline{G}^\circ \left[O_2(g) \right]$$

$$= 2\left(-110.5 \text{ kJ mol}^{-1} \right) - 2 \left(-34.6 \text{ kJ mol}^{-1} \right) - 0 \text{ kJ mol}^{-1}$$

$$= -151.8 \text{ kJ mol}^{-1}$$

When 1 mole of NO_2^- is oxidized to 1 mole of NO_3^-,

$$\Delta_r G^\circ = \frac{-151.8 \text{ kJ mol}^{-1}}{2} = -75.9 \text{ kJ mol}^{-1}$$

4.34   This problem involves the synthesis of diamond from graphite:

$$C(\text{graphite}) \rightarrow C(\text{diamond})$$

(a) Calculate the values of $\Delta_r H^\circ$ and $\Delta_r S^\circ$ for the reaction. Will the conversion be favored at $25°$ C or any other temperature? **(b)** From density measurements, the molar volume of graphite is found to be 2.1 cm^3 greater than that of diamond. Can the conversion of graphite to diamond be brought about at $25°$ C by applying pressure on graphite? If so, estimate the pressure at which the process becomes spontaneous. [*Hint*: Starting from Equation 4.32, derive the equation $\Delta_r G_2 - \Delta_r G_1 = \left(\overline{V}_{\text{diamond}} - \overline{V}_{\text{graphite}} \right) \Delta P$ for a constant-temperature process. Next, calculate the ΔP value that would lead to the necessary decrease in Gibbs energy.]

(a)
$$\Delta_r H^\circ = \Delta_f \overline{H}^\circ \, [C(\text{diamond})] - \Delta_f \overline{H}^\circ \, [C(\text{graphite})]$$

$$= 1.90 \text{ kJ mol}^{-1} - 0 \text{ kJ mol}^{-1}$$

$$= 1.90 \text{ kJ mol}^{-1}$$

$$\Delta_r S^\circ = \overline{S}^\circ \, [C(\text{diamond})] - \overline{S}^\circ \, [C(\text{graphite})]$$

$$= 2.4 \text{ J K}^{-1} \text{mol}^{-1} - 5.7 \text{ J K}^{-1} \text{mol}^{-1}$$

$$= -3.3 \text{ J K}^{-1} \text{mol}^{-1}$$

At 25° C,

$$\Delta_r G^\circ = \Delta_r H^\circ - (298 \text{ K}) \, \Delta_r S^\circ$$

$$= 1.90 \text{ kJ mol}^{-1} - (298 \text{ K}) \left(-3.3 \times 10^{-3} \text{ kJ K}^{-1} \text{mol}^{-1} \right)$$

$$= 2.883 \text{ kJ mol}^{-1} = 2.88 \text{ kJ mol}^{-1}$$

Therefore, the conversion from graphite to diamond is not spontaneous at 25° C and when both are in their standard states. In fact, because $\Delta_r H^\circ$ is positive and $T \Delta_r S^\circ$ is negative, $\Delta_r G^\circ$ can never be negative. Thus, the conversion will not be spontaneous at any other temperature.

(b) The integration of Equation 4.32

$$\left(\frac{\partial G}{\partial P} \right)_T = V$$

gives

$$G_2 = G_1 + V \left(P_2 - P_1 \right)$$

where G_1 and G_2 are the Gibbs energies at P_1 and P_2, respectively. Using molar quantities and $\Delta P = P_2 - P_1$, the equation becomes

$$\overline{G}_2 = \overline{G}_1 + \overline{V} \Delta P$$

Apply this equation to graphite and diamond,

$$\overline{G}_2 \, [\text{graphite}] = \overline{G}_1 \, [\text{graphite}] + \overline{V} \, [\text{graphite}] \, \Delta P$$

$$\overline{G}_2 \, [\text{diamond}] = \overline{G}_1 \, [\text{diamond}] + \overline{V} \, [\text{diamond}] \, \Delta P$$

These two equations are combined to relate the values of $\Delta_r G$ for the conversion of graphite to diamond at two different pressures:

$$\Delta_r G_2 = \overline{G}_2 \, [\text{diamond}] - \overline{G}_2 \, [\text{graphite}]$$

$$= \left\{ \overline{G}_1 \, [\text{diamond}] + \overline{V} \, [\text{diamond}] \, \Delta P \right\} - \left\{ \overline{G}_1 \, [\text{graphite}] + \overline{V} \, [\text{graphite}] \, \Delta P \right\}$$

$$= \Delta_r G_1 + \left\{ \overline{V} \, [\text{diamond}] - \overline{V} \, [\text{graphite}] \right\} \Delta P$$

If $P_1 = 1$ bar, then $\Delta_r G_1 = \Delta_r G^\circ = 2.883 \text{ kJ mol}^{-1}$ at 25° C. $\Delta_r G_2$ becomes

$$\Delta_r G_2 = 2.883 \text{ kJ mol}^{-1} + \left(-2.1 \text{ cm}^3 \text{ mol}^{-1} \right) \Delta P \left(\frac{1 \text{ L}}{1000 \text{ cm}^3} \right) \left(\frac{1 \text{ atm}}{1.013 \text{ bar}} \right) \left(\frac{101.3 \text{ J}}{1 \text{ L atm}} \right) \left(\frac{1 \text{ kJ}}{1000 \text{ J}} \right)$$

$$= 2.883 \text{ kJ mol}^{-1} - 2.1 \times 10^{-4} \Delta P \text{ kJ bar}^{-1} \text{mol}^{-1}$$

If the process is spontaneous, then

$$\Delta_r G_2 = 2.883 \text{ kJ mol}^{-1} - 2.1 \times 10^{-4} \Delta P \text{ kJ bar}^{-1} \text{ mol}^{-1} < 0$$

$$2.883 \text{ kJ mol}^{-1} < 2.1 \times 10^{-4} \Delta P \text{ kJ bar}^{-1} \text{ mol}^{-1}$$

$$\Delta P > \frac{2.883 \text{ kJ mol}^{-1}}{2.1 \times 10^{-4} \text{ kJ bar}^{-1} \text{ mol}^{-1}} = 1.4 \times 10^4 \text{ bar}$$

Therefore, at a very high pressure, $P_2 = \Delta P + P_1 = 1.4 \times 10^4 \text{ bar} + 1 \text{ bar} = 1.4 \times 10^4 \text{ bar}$, the conversion from graphite to diamond is spontaneous.

4.36 Predict the signs of ΔH, ΔS, and ΔG of the system for the following processes at 1 atm: **(a)** ammonia melts at $-60°$ C, **(b)** ammonia melts at $-77.7°$ C, and **(c)** ammonia melts at $-100°$ C. (The normal melting point of ammonia is $-77.7°$ C.)

Since melting is an endothermic process, ΔH is positive in all three cases. A substance in the liquid phase is more disordered than in the solid phase. Therefore, ΔS is positive in all three cases.

When the temperature is above the melting point, the melting process is spontaneous, that is, ΔG is negative. When the temperature is at the melting point, the melting process is at equilibrium, that is, ΔG is zero. When the temperature is below the melting point, the melting process is not spontaneous, that is, ΔG is negative.

In summary,

(a) ΔH: +, ΔS: +, ΔG: −

(b) ΔH: +, ΔS: +, ΔG: 0

(c) ΔH: +, ΔS: +, ΔG: +

4.38 A student looked up the $\Delta_f \overline{G}°$, $\Delta_f \overline{H}°$, and $\overline{S}°$ values for CO_2 in Appendix 2. Plugging these values into Equation 4.21, he found that $\Delta_f \overline{G}° \neq \Delta_f \overline{H}° - T\overline{S}°$ at 298 K. What is wrong with his approach?

The equation $\Delta G° = \Delta H° - T\Delta S°$ applies to a process. If the process is a formation reaction, such as the formation of CO_2 [C(graphite) + $O_2(g) \rightarrow CO_2(g)$], then

$$\Delta_r G° = \Delta_f G° = -394.4 \text{ kJ mol}^{-1}$$

$$\Delta_r H° = \Delta_f H° = -393.5 \text{ kJ mol}^{-1}$$

$$\Delta_r S° = \overline{S}° \left[CO_2(g)\right] - \overline{S}° \left[C(\text{graphite})\right] - \overline{S}° \left[O_2(g)\right]$$

$$= 213.6 \text{ J K}^{-1} \text{ mol}^{-1} - 5.7 \text{ J K}^{-1} \text{ mol}^{-1} - 205.0 \text{ J K}^{-1} \text{ mol}^{-1}$$

$$= 2.9 \text{ J K}^{-1} \text{ mol}^{-1}$$

Note that $\Delta_r S°$ is not the same as $\overline{S}° \left[CO_2(g)\right]$ as stated in the question. At 298 K,

$$\Delta_r H^\circ - T\Delta_r S^\circ = -393.5 \text{ kJ mol}^{-1} - (298 \text{ K}) \left(2.9 \times 10^{-3} \text{ kJ K}^{-1} \text{mol}^{-1}\right)$$

$$= -394.4 \text{ kJ mol}^{-1}$$

which is the same as $\Delta_r G^\circ$.

4.40 A certain reaction is known to have a $\Delta_r G^\circ$ value of -122 kJ. Will the reaction necessarily occur if the reactants are mixed together?

No, for two reasons. First, the reaction rate can be very slow. Secondly, even if the rate is rapid enough for reaction to occur, the value of $\Delta_r G$ for the reaction depends on the concentrations of reactants and products and will be different from the value $\Delta_r G^\circ$ that is appropriate when all species are present in their standard pressure or concentration.

4.42 The pressure exerted on ice by a 60.0-kg skater is about 300 atm. Calculate the depression in freezing point. The molar volumes are $\overline{V}_L = 0.0180 \text{ L mol}^{-1}$ and $\overline{V}_S = 0.0196 \text{ L mol}^{-1}$.

The depression of freezing point can be obtained from the slope of the S–L curve in the phase diagram.

$$\frac{dP}{dT} = \frac{\Delta\overline{H}}{T\Delta\overline{V}} = \frac{\left(6.01 \times 10^3 \text{ J mol}^{-1}\right)\left(\frac{1 \text{ L atm}}{101.3 \text{ J}}\right)}{(273.2 \text{ K})\left(0.0180 \text{ L mol}^{-1} - 0.0196 \text{ L mol}^{-1}\right)}$$

$$= -135.7 \text{ atm K}^{-1}$$

$$dP = -135.7 \text{ atm K}^{-1} dT$$

$$\int_{1 \text{ atm}}^{300 \text{ atm}} dP = -135.7 \text{ atm K}^{-1} \int_{273.15 \text{ K}}^{T \text{ K}} dT$$

$$299 \text{ atm} = -135.7 \text{ atm K}^{-1}\Delta T$$

$$\Delta T = \frac{299 \text{ atm}}{-135.7 \text{ atm K}^{-1}} = -2.20 \text{ K}$$

Therefore, the freezing point is depressed by 2.20 K when 300 atm is applied to ice. In other words, the freezing point at this pressure is $273.15 \text{ K} - 2.20 \text{ K} = 270.95 \text{ K}$ or $-2.20\,^\circ\text{C}$.

4.44 Use the phase diagram of water (Figure 4.14) to predict the dependence of the freezing and boiling points of water on pressure.

The freezing point decreases with increasing pressure (because the S–L curve has a negative slope) whereas the boiling point increases with increasing pressure (because the L–V curve has a positive slope).

4.46 Below is a rough sketch of the phase diagram of carbon. (a) How many triple points are there, and what are the phases that can coexist at each triple point? (b) Which has a higher density, graphite or diamond? (c) Synthetic diamond can be made from graphite. Using the phase diagram, how would you go about making diamond?

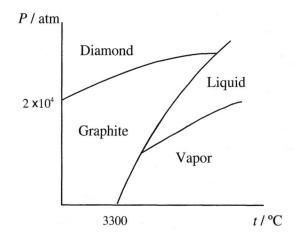

(a) There are two triple points. At one triple point, diamond, graphite and liquid carbon coexist. At the other, graphite, liquid carbon, and gaseous carbon coexist.

(b) According to the Clapeyron equation, $\frac{dP}{dT} = \frac{\Delta \overline{H}}{T\Delta \overline{V}}$. From the phase diagram, the graphite-diamond curve has positive slope for most of its length. From Problem 4.34, the standard molar enthalpy change for C(graphite) $\rightarrow$ C(diamond), $\Delta \overline{H}^\circ$, is also positive. This implies that $\Delta \overline{V}$ is positive, or that diamond has a greater molar volume than graphite. Consequently, one would conclude that diamond is less dense than graphite. In fact this conclusion is in error, and diamond is the denser phase. The problem is that by using $\Delta \overline{H}^\circ$, the assumption is made that ΔH is independent of pressure. Indeed, the evidence suggests that it changes sign and is negative at those pressures where graphite and diamond are in equilibrium.

(c) The phase diagram indicates that high pressures are required to convert graphite to diamond.

4.48   The plot in Figure 4.13 is no longer linear at high temperatures. Explain.

The plot is no longer linear at high temperatures means that the slope is not constant over a large temperature range. This is because $\Delta_{vap}\overline{H}$, which is related to the slope, remains constant only over a limited temperature range.

4.50   The normal boiling point of ethanol is 78.3° C, and its molar enthalpy of vaporization is 39.3 kJ mol^{-1}. What is its vapor pressure at 30° C?

$$\ln \frac{P_2}{P_1} = -\frac{\Delta_{vap}\overline{H}}{R}\left(\frac{1}{T_2} - \frac{1}{T_1}\right)$$

$$\ln \frac{P_2}{1\ \text{atm}} = -\frac{39.3 \times 10^3\ \text{J mol}^{-1}}{8.314\ \text{J K}^{-1}\ \text{mol}^{-1}}\left(\frac{1}{303.2\ \text{K}} - \frac{1}{351.5\ \text{K}}\right) = -2.142$$

$$P_2 = (1\ \text{atm})\, e^{-2.142} = 0.117\ \text{atm} = 88.9\ \text{torr}$$

4.52 State the condition(s) under which the following equations can be applied: (a) $\Delta S = \Delta H/T$, (b) $S_0 = 0$, (c) $dS = C_P dT/T$, and (d) $dS = dq/T$.

(a) Constant pressure and temperature, reversible process.

(b) Absolute zero, no residual entropy and a pure, crystalline substance.

(c) Constant pressure (Note that when this form is integrated, one must either include the explicit temperature dependence of C_P or assume it to be temperature independent.)

(d) reversible process

4.54 Calculate the entropy change when neon at 25° C and 1.0 atm in a container of volume 0.780 L is allowed to expand to 1.25 L and is simultaneously heated to 85° C. Assume ideal behavior. (*Hint*: Because S is a state function, you can first calculate the value of ΔS for expansion and then calculate the value of ΔS for heating at constant final volume.)

The number of moles of Ne can be determined using the initial conditions and the ideal gas law.

$$n = \frac{(1.0 \text{ atm}) (0.780 \text{ L})}{(0.08206 \text{ L atm K}^{-1} \text{ mol}^{-1}) (298 \text{ K})} = 3.19 \times 10^{-2} \text{ mol}$$

The problem can be solved by breaking down the process into 2 steps: (1) isothermal expansion from 0.780 L at 1.0 atm to 1.25 L. The temperature is kept at 20° C; (2) heating at constant volume (1.25 L) from 25° C to 85° C. The entropy changes for these two steps, ΔS_1 and ΔS_2 are

$$\Delta S_1 = nR \ln \frac{V_2}{V_1} = \left(3.19 \times 10^{-2} \text{ mol}\right) \left(8.314 \text{ J K}^{-1} \text{mol}^{-1}\right) \ln \frac{1.25 \text{ L}}{0.780 \text{ L}} = 0.125 \text{ J K}^{-1}$$

$$\Delta S_2 = C_V \ln \frac{T_2}{T_1} = \frac{3}{2} nR \ln \frac{T_2}{T_1} = \frac{3}{2} \left(3.19 \times 10^{-2} \text{ mol}\right) \left(8.314 \text{ J K}^{-1} \text{mol}^{-1}\right) \ln \frac{358 \text{ K}}{298 \text{ K}}$$

$$= 7.30 \times 10^{-2} \text{ J K}^{-1}$$

The entropy change for the entire process is

$$\Delta S = \Delta S_1 + \Delta S_2 = 0.125 \text{ J K}^{-1} + 7.30 \times 10^{-2} \text{ J K}^{-1} = 0.20 \text{ J K}^{-1}$$

4.56 One mole of an ideal monatomic gas is compressed from 2.0 atm to 6.0 atm while being cooled from 400 K to 300 K. Calculate the values of ΔU, ΔH, and ΔS for the process.

The thermodynamic quantities can be readily calculated by breaking down the process into 2 steps: (1) isothermal compression at 400 K from 2.0 atm to 6.0 atm; (2) cooling at constant pressure (6.0 atm) from 400 K to 300 K.

Step 1

For an ideal gas undergoing an isothermal process, $\Delta U = 0$ and $\Delta H = 0$.

$$\Delta S = nR \ln \frac{V_2}{V_1} = nR \ln \frac{P_1}{P_2} = (1 \text{ mol}) \left(8.314 \text{ J K}^{-1} \text{mol}^{-1}\right) \ln \frac{2.0 \text{ atm}}{6.0 \text{ atm}} = -9.13 \text{ J K}^{-1}$$

Step 2

$$\Delta H = C_P \Delta T = \frac{5}{2} nR\Delta T = \frac{5}{2}(1 \text{ mol})\left(8.314 \text{ J K}^{-1}\text{mol}^{-1}\right)(300 \text{ K} - 400 \text{ K}) = -2.079 \times 10^3 \text{ J}$$

$$\Delta U = \Delta H - nR\Delta T = -2.079 \times 10^3 \text{ J} - (1 \text{ mol})\left(8.314 \text{ J K}^{-1}\text{mol}^{-1}\right)(300 \text{ K} - 400 \text{ K})$$

$$= -1.248 \times 10^3 \text{ J}$$

$$\Delta S = C_P \ln\frac{T_2}{T_1} = \frac{5}{2} nR \ln\frac{T_2}{T_1} = \frac{5}{2}(1 \text{ mol})\left(8.314 \text{ J K}^{-1}\text{mol}^{-1}\right) \ln\frac{300 \text{ K}}{400 \text{ K}} = -5.979 \text{ J K}^{-1}$$

For the entire process,

$$\Delta H = 0 \text{ J} - 2.079 \times 10^3 \text{ J} = -2.08 \times 10^3 \text{ J}$$

$$\Delta U = 0 \text{ J} - 1.248 \times 10^3 \text{ J} = -1.25 \times 10^3 \text{ J}$$

$$\Delta S = -9.13 \text{ J K}^{-1} - 5.979 \text{ J K}^{-1} = -15.1 \text{ J K}^{-1}$$

4.58 Use the following data to determine the normal boiling point, in kelvins, of mercury. What assumptions must you make to do the calculation?

$$Hg(l): \quad \Delta_f \overline{H}^\circ = 0 \text{ (by definition)}$$

$$\overline{S}^\circ = 77.4 \text{ J K}^{-1}\text{mol}^{-1}$$

$$Hg(g): \quad \Delta_f \overline{H}^\circ = 60.78 \text{ kJ mol}^{-1}$$

$$\overline{S}^\circ = 174.7 \text{ J K}^{-1}\text{mol}^{-1}$$

The normal boiling point of mercury, T_b, is related to $\Delta_{vap}S^\circ$ and $\Delta_{vap}H^\circ$:

$$\Delta_{vap}S^\circ = \frac{\Delta_{vap}H^\circ}{T_b}$$

For 1 mole of mercury, the entropy and enthalpy of vaporization are

$$\Delta_{vap}H^\circ = 60.78 \text{ kJ} - 0 \text{ kJ} = 60.78 \text{ kJ}$$

$$\Delta_{vap}S^\circ = 174.7 \text{ J K}^{-1} - 77.4 \text{ J K}^{-1} = 97.3 \text{ J K}^{-1}$$

Therefore,

$$T_b = \frac{\Delta_{vap}H^\circ}{\Delta_{vap}S^\circ} = \frac{60.78 \times 10^3 \text{ J}}{97.3 \text{ J K}^{-1}} = 625 \text{ K} = 352^\circ \text{ C}$$

The assumptions made in this calculation are that the values of $\Delta_f \overline{H}^\circ$ and $\overline{S}^\circ$ are temperature independent. These assumptions are quite good because the calculated boiling point of mercury is very close to the actual value of 356.6° C.

4.60 Give a detailed example of each of the following, with an explanation: **(a)** a thermodynamically spontaneous process; **(b)** a process that would violate the first law of thermodynamics; **(c)** a process that would violate the second law of thermodynamics; **(d)** an irreversible process; and **(e)** an equilibrium process.

(a) An ice cube melting in a glass of water at 20° C.

(b) A perpetual motion machine of the first kind, such as a rotating flywheel that drives a generator, the output of which is used to keep the flywheel rotating at a constant speed and also to lift a weight.

(c) A perfect air conditioner; it extracts heat from the room and warms the outside without using any energy to do so. (This does not violate the first law, since the energy deposited outside is exactly equal to that removed from inside.)

(d) Same as part **(a)**, an ice cube melting in a glass of water at 20° C.

(e) Water and ice in a closed system at 0° C and 1 atm pressure.

4.62 Superheated water is water heated above 100° C without boiling. As for supercooled water (see Example 4.4), superheated water is thermodynamically unstable. Calculate the values of ΔS_{sys}, ΔS_{surr}, and ΔS_{univ} when 1.5 moles of superheated water at 110° C and 1.0 atm are converted to steam at the same temperature and pressure. (The molar enthalpy of vaporization of water is 40.79 kJ mol^{-1}, and the molar heat capacities of water and steam in the temperature range 100–110° C are 75.5 J K^{-1} mol^{-1} and 34.4 J K^{-1} mol^{-1}, respectively.)

The entire process can be broken down into 3 steps: (1) cooling of $H_2O(l)$ from 110° C to 100° C; (2) supplying heat to system to effect the phase transformation of H_2O from liquid to gas at 100° C; and (3) heating $H_2O(g)$ from 100° C to 110° C.

Step 1

$$H_2O(l, 110° C) \rightarrow H_2O(l, 100° C)$$

$$\Delta S_1 = C_P \ln \frac{373 \text{ K}}{383 \text{ K}} = (1.5 \text{ mol}) \left(75.5 \text{ J K}^{-1} \text{mol}^{-1}\right) \ln \frac{373 \text{ K}}{383 \text{ K}} = -3.00 \text{ J K}^{-1}$$

$$q_{surr,1} = -q_1 = -C_P \Delta T = -(1.5 \text{ mol}) \left(75.5 \text{ J K}^{-1} \text{mol}^{-1}\right)(373 \text{ K} - 383 \text{ K}) = 1.13 \times 10^3 \text{J}$$

Step 2

$$H_2O(l, 100° C) \rightarrow H_2O(g, 100° C)$$

$$\Delta S_2 = \frac{\Delta_{vap} H}{T_b} = \frac{(1.5 \text{ mol}) \left(40.79 \times 10^3 \text{ J mol}^{-1}\right)}{373 \text{ K}} = 164 \text{ J K}^{-1}$$

$$q_{surr,2} = -q_2 = -\Delta_{vap} H = -(1.5 \text{ mol}) \left(40.79 \times 10^3 \text{ J mol}^{-1}\right) = -6.12 \times 10^4 \text{ J}$$

Step 3

$$H_2O(g, 100° C) \to H_2O(g, 110° C)$$

$$\Delta S_3 = C_P \ln \frac{383 \text{ K}}{373 \text{ K}} = (1.5 \text{ mol}) \left(34.4 \text{ J K}^{-1} \text{mol}^{-1}\right) \ln \frac{383 \text{ K}}{373 \text{ K}} = 1.37 \text{ J K}^{-1}$$

$$q_{\text{surr},3} = -q_3 = -C_P \Delta T = -(1.5 \text{ mol}) \left(34.4 \text{ J K}^{-1} \text{mol}^{-1}\right) (383 \text{ K} - 373 \text{ K}) = -516 \text{ J}$$

Therefore, for the entire process,

$$\Delta S_{\text{sys}} = -3.00 \text{ J K}^{-1} + 164 \text{ J K}^{-1} + 1.37 \text{ J K}^{-1} = 162 \text{ J K}^{-1}$$

$$q_{\text{surr}} = 1.13 \times 10^3 \text{ J} - 6.12 \times 10^4 \text{ J} - 516 \text{ J} = -6.06 \times 10^4 \text{ J}$$

$$\Delta S_{\text{surr}} = \frac{q_{\text{surr}}}{T_{\text{surr}}} = \frac{-6.06 \times 10^4 \text{ J}}{383 \text{ K}} = -158 \text{ J K}^{-1}$$

$$\Delta S_{\text{univ}} = 162 \text{ J K}^{-1} - 158 \text{ J K}^{-1} = 4 \text{ J K}^{-1}$$

$\Delta S_{\text{univ}} > 0$ confirms the expectation that superheated water vaporizes spontaneously.

4.64 Give the conditions under which each of the following equations may be applied: (a) $dA \leq 0$ (for equilibrium and spontaneity), (b) $dG \leq 0$ (for equilibrium and spontaneity), (c) $\ln \frac{P_2}{P_1} = \frac{\Delta \overline{H}}{R} \frac{(T_2 - T_1)}{T_1 T_2}$, and (d) $\Delta G = nRT \ln \frac{P_2}{P_1}$.

(a) Constant volume and temperature, (b) constant pressure and temperature, (c) $\Delta_{\text{vap}} \overline{H}$ independent of temperature, $\overline{V}_{\text{vap}} \gg \overline{V}_{\text{liquid}}$ (or $\overline{V}_{\text{solid}}$), ideal gas behavior, (d) constant temperature, ideal gas behavior.

4.66 Protein molecules are polypeptide chains made up of amino acids. In their physiologically functioning or native state, these chains fold in a unique manner such that the nonpolar groups of the amino acids are usually buried in the interior region of the proteins, where there is little or no contact with water. When a protein denatures, the chain unfolds so that these nonpolar groups are exposed to water. A useful estimate of the changes of the thermodynamic quantities as a result of denaturation is to consider the transfer of a hydrocarbon such as methane (a nonpolar substance) from an inert solvent (such as benzene or carbon tetrachloride) to the aqueous environment:

(a) $CH_4(\text{inert solvent}) \to CH_4(g)$

(b) $CH_4(g) \to CH_4(aq)$

If the values of $\Delta H°$ and $\Delta G°$ are approximately 2.0 kJ mol^{-1} and -14.5 kJ mol^{-1}, respectively, for (a) and -13.5 kJ mol^{-1} and 26.5 kJ mol^{-1}, respectively, for (b), calculate the values of $\Delta H°$ and $\Delta G°$ for the transfer of 1 mole of CH_4 according to the equation

$$CH_4(\text{inert solvent}) \to CH_4(aq)$$

Comment on your results. Assume $T = 298$ K.

For the process $CH_4(\text{inert solvent}) \to CH_4(aq)$

$$\Delta H^\circ = 2.0 \text{ kJ mol}^{-1} - 13.5 \text{ kJ mol}^{-1} = -11.5 \text{ kJ mol}^{-1}$$

$$\Delta G^\circ = -14.5 \text{ kJ mol}^{-1} + 26.5 \text{ kJ mol}^{-1} = 12.0 \text{ kJ mol}^{-1}$$

Since $\Delta G^\circ > 0$, the process is not spontaneous when the reactant and product are in their standard states. Furthermore,

$$\Delta S^\circ = \frac{\Delta H^\circ - \Delta G^\circ}{T} = \frac{-11.5 \text{ kJ mol}^{-1} - 12.0 \text{ kJ mol}^{-1}}{298 \text{ K}} \left(\frac{1000 \text{ J}}{1 \text{ kJ}} \right) = -78.9 \text{ J K}^{-1} \text{mol}^{-1}$$

This negative value of ΔS° indicates that there is a increase in order when CH_4 is dissolved in H_2O. This is a result of order imposed on the solvent (water) molecules, due to the special arrangement of water molecules around each CH_4 molecule (see Section 13.6).

4.68 A rubber band under tension will contract when heated. Explain.

When the rubber band is heated, ΔH is positive. If a process occurs, ΔG must be negative.

$$\Delta G = \Delta H - T \Delta S$$

$$\Delta S = \frac{\Delta H - \Delta G}{T}$$

The expression on the right hand side of the last expression is positive, that is, ΔS must also be positive for a process to occur. The process that can effect such a ΔS is the contraction of the rubber band.

4.70 A sample of supercooled water freezes at -10° C. What are the signs of ΔH, ΔS, and ΔG for this process? All the changes refer to the system.

ΔG: $-$ because the process is spontaneous (constant temperature given and constant pressure assumed).

ΔH: $-$ because freezing is an exothermic process.

ΔS: $-$ because ice is more ordered than liquid water.

4.72 A chemist has synthesized a hydrocarbon compound (C_xH_y). Briefly describe what measurements are needed to determine the values of $\Delta_f \overline{H}^\circ$, $\overline{S}^\circ$, and $\Delta_f \overline{G}^\circ$ of the compound.

A determination of the enthalpy of combustion $[C_xH_y + (x + \frac{y}{4})O_2(g) \rightarrow xCO_2(g) + \frac{y}{2}H_2O(l)]$ using a calorimeter will enable a calculation of $\Delta_f \overline{H}^\circ$ from this measurement and the known $\Delta_f \overline{H}^\circ$'s for O_2, CO_2 and H_2O.

$\overline{S}^\circ$ may be found via a determination of the third-law entropy from 0 K to 298 K. This assumes no residual entropy at 0 K.

Once $\overline{S}^\circ$ is known for the compound, it is used together with the known values of $\overline{S}^\circ$ for C(graphite) and $H_2(g)$ to calculate $\Delta_r S^\circ$ for the reaction xC(graphite) $+ \frac{y}{2}H_2(g) \rightarrow C_xH_y$. $\Delta_f \overline{G}^\circ$ for the compound is determined via $\Delta_f \overline{G}^\circ = \Delta_f \overline{H}^\circ - T \Delta_r S^\circ$.

4.74 A person heated water in a closed bottle in a microwave oven for tea. After removing the bottle from the oven, she added a tea bag to the hot water. To her surprise, the water started to boil violently. Explain what happened.

The water was superheated. The closed bottle allowed the pressure to become greater than 1 atm during the heating. Although thermodynamically unstable, superheated water will not boil even when exposed to a pressure of about 1 atm. Any mechanical disturbance, such as shaking, will cause it to boil. Adding the tea bag acts like adding boiling chips, which facilitates the boiling action.

4.76 The molar entropy of argon (Ar) is given by

$$\overline{S}^{o} = (36.4 + 20.8 \ln T) \ \text{J K}^{-1} \text{mol}^{-1}$$

Calculate the change in Gibbs energy when 1.0 mole of Ar is heated at constant pressure from $20°$ C to $60°$ C. (*Hint*: Use the relation $\int \ln x \, dx = x \ln x - x$.)

ΔG^{o} can be evaluated by using the relation

$$\left(\frac{\partial G}{\partial T}\right)_{P} = -S$$

At constant pressure,

$$dG = -S dT$$

$$\int_{G_1}^{G_2} dG = -\int_{T_1}^{T_2} S dT$$

$$\Delta G = -\int_{293.2 \text{ K}}^{333.2 \text{ K}} (1 \text{ mol}) \left[(36.4 + 20.8 \ln T) \ \text{J K}^{-1} \text{mol}^{-1}\right] dT$$

$$= -[36.4T + 20.8 (T \ln T - T)]_{293.2}^{333.2} \text{ J}$$

$$= -\{[36.4 (333.2) + 20.8 (333.2 \ln 333.2 - 333.2)]$$

$$- [36.4 (293.2) + 20.8 (293.2 \ln 293.2 - 293.2)]\} \text{ J}$$

$$= -6.24 \times 10^3 \text{ J}$$

4.78 Comment on the analogy sometimes used to relate a student's dormitory room becoming disorderly and untidy to an increase in entropy.

The analogy is inappropriate. Entropy is a measure of the dispersal of molecules among available energy levels. The entropy of the room is the same whether it is tidy or not.

4.80 In a DSC experiment (see p. 62), the melting temperature (T_m) of a certain protein is found to be $46°$ C and the enthalpy of denaturation is 382 kJ mol^{-1}. Estimate the entropy of denaturation assuming that the denaturation is a two-state process; that is, native protein $\rightleftharpoons$ denatured protein. The single polypeptide protein chain has 122 amino acids. Calculate the entropy of denaturation per amino acid. Comment on your result.

$\Delta G = \Delta H - T \Delta S$. Since at the melting temperature, $\Delta G = 0$, it follows that

$$\Delta S = \frac{\Delta H}{T_{\mathrm{m}}} = \frac{382 \times 10^3 \,\mathrm{J\,mol^{-1}}}{(46 + 273.15)\,\mathrm{K}} = 1.20 \times 10^3 \,\mathrm{J\,K^{-1}\,mol^{-1}}$$

With 122 amino acids in the protein, the entropy change per (mole) amino acid is

$$\frac{1.20 \times 10^3 \,\mathrm{J\,K^{-1}\,mol^{-1}}}{122} = 9.84 \,\mathrm{J\,K^{-1}\,mol\,amino\,acid^{-1}}$$

This is relatively small and may be compared with $\Delta_{\mathrm{fus}} S = 22 \,\mathrm{J\,K^{-1}\,mol^{-1}}$ for water.

CHAPTER 5

Solutions

PROBLEMS AND SOLUTIONS

5.2 What is the molarity of a 2.12 mol kg^{-1} aqueous sulfuric acid solution? The density of this solution is 1.30 g cm^{-3}.

To find the molarity of the solution, the number of moles of solute and the volume of solution in a sample must be determined. Assume 1 kg of water is present in the solution.

$$\text{Number of moles of } H_2SO_4 = 2.12 \text{ mol}$$

$$\text{Mass of solution} = \text{mass of water} + \text{mass of } H_2SO_4$$

$$= 1000 \text{ g} + (2.12 \text{ mol } H_2SO_4) \left(\frac{98.09 \text{ g}}{1 \text{ mol } H_2SO_4} \right) = 1208.0 \text{ g}$$

$$\text{Volume of solution} = \frac{1208.0 \text{ g}}{(1.30 \text{ g cm}^{-3}) \left(\frac{1000 \text{ cm}^3}{1 \text{ L}} \right)} = 0.9292 \text{ L}$$

$$\text{Molarity of solution} = \frac{2.12 \text{ mol}}{0.9292 \text{ L}} = 2.28 \text{ } M$$

5.4 The concentrated sulfuric acid we use in the laboratory is 98.0% sulfuric acid by weight. Calculate the molality and molarity of concentrated sulfuric acid if the density of the solution is 1.83 g cm^{-3}.

Assume 100 g of solution is present. The solution contains 98.0 g H_2SO_4 and 2.0 g H_2O.

$$\text{Number of moles of } H_2SO_4 = (98.0 \text{ g}) \left(\frac{1 \text{ mol}}{98.09 \text{ g}} \right) = 0.9991 \text{ mol}$$

The molality of the solution is the ratio between the number of moles of solute and the mass of solvent:

$$\text{Molality of solution} = \frac{0.9991 \text{ mol}}{2.0 \times 10^{-3} \text{ kg}} = 5.0 \times 10^2 \text{ } m$$

The molarity of the solution is the ratio between the number of moles of solute and the volume of the solution.

$$\text{Volume of solution} = \frac{100 \text{ g}}{\left(1.83 \text{ g cm}^{-3}\right)\left(\frac{1000 \text{ cm}^3}{1 \text{ L}}\right)} = 5.464 \times 10^{-2} \text{ L}$$

$$\text{Molarity of solution} = \frac{0.9991 \text{ mol}}{5.464 \times 10^{-2} \text{ L}} = 18.3 \, M$$

5.6   For dilute aqueous solutions in which the density of the solution is roughly equal to that of the pure solvent, the molarity of the solution is equal to its molality. Show that this statement is correct for a 0.010 M aqueous urea [$(NH_2)_2CO$] solution.

To convert molarity to molality, the volume of the solution has to be converted to the mass of solvent. Assume 1 L of solution is present.

$$\text{Number of moles of urea} = 0.010 \text{ mol}$$

$$\text{Mass of solvent} = \text{mass of solution} - \text{ mass of solute}$$

$$= \left(1000 \text{ cm}^3\right)\left(1.00 \text{ g cm}^{-3}\right) - (0.010 \text{ mol urea})\left(\frac{60.06 \text{ g}}{1 \text{ mol urea}}\right)$$

$$= 999.4 \text{ g} = 0.9994 \text{ kg}$$

$$\text{Molality of solution} = \frac{0.010 \text{ mol}}{0.9994 \text{ kg}} = 0.010 \, m$$

Therefore, for a dilute aqueous solution, such as 0.010 M urea, its molality is numerically the same as its molarity.

5.8   The strength of alcoholic beverages is usually described in terms of "proof," which is defined as twice the percentage by volume of ethanol. Calculate the number of grams of alcohol in 2 quarts of 75-proof gin. What is the molality of the gin? (The density of ethanol is 0.80 g cm^{-3}; 1 quart = 0.946 L.)

Since mass of a liquid is proportional to its volume, the % by weight of ethanol is the same as % by volume of ethanol, which is half the proof, or 37.5%. To find the molality of the gin, the number of moles of ethanol and the mass of water in a quantity of gin have to be calculated. In 2 quarts of gin,

$$\text{Volume of ethanol} = (37.5\%)\left(2 \text{ quarts}\right)\left(\frac{0.946 \text{ L}}{1 \text{ quart}}\right) = 0.7095 \text{ L}$$

$$\text{Mass of ethanol} = \left(0.7095 \times 10^3 \text{ cm}^3\right)\left(0.80 \text{ g cm}^{-3}\right) = 568 \text{ g} = 5.7 \times 10^2 \text{ g}$$

$$\text{Number of moles of ethanol} = \left(568 \text{ g}\right)\left(\frac{1 \text{ mol}}{46.07 \text{ g}}\right) = 12.3 \text{ mol}$$

To evaluate the mass of water in 2 quarts of gin, two assumptions have to be made: (1) the volumes of ethanol and water are additive, and (2) the density of water is 1 g cm^{-3}. The second assumption is not bad, but the first is not particularly good. Indeed, there is a significant nonideality in water-ethanol solutions.

Volume of water in 2 quarts of gin $=$ volume of gin $-$ volume of ethanol

$$= (2 \text{ quarts}) \left(\frac{0.946 \text{ L}}{1 \text{ quart}} \right) - 0.7095 \text{ L} = 1.1825 \text{ L}$$

$$\text{Mass of water in 2 quarts of gin} = \left(1.1825 \times 10^3 \text{ cm}^3 \right) \left(1 \text{ g cm}^{-3} \right)$$

$$= 1.1825 \times 10^3 \text{ g} = 1.1825 \text{ kg}$$

Therefore,

$$\text{Molality of the gin} = \frac{12.3 \text{ mol}}{1.1825 \text{ kg}} = 10 \ m$$

5.10   Calculate the changes in entropy for the following processes: **(a)** mixing of 1 mole of nitrogen and 1 mole of oxygen, and **(b)** mixing of 2 moles of argon, 1 mole of helium, and 3 moles of hydrogen. Both **(a)** and **(b)** are carried out under conditions of constant temperature (298 K) and constant pressure. Assume ideal behavior.

(a) $\Delta_{mix}S = -nR \left(x_{N_2} \ln x_{N_2} + x_{O_2} \ln x_{O_2} \right)$

$$= -(2 \text{ mol}) \left(8.314 \text{ J K}^{-1} \text{mol}^{-1} \right) \left(\frac{1}{2} \ln \frac{1}{2} + \frac{1}{2} \ln \frac{1}{2} \right) = 11.53 \text{ J K}^{-1}$$

(b) $\Delta_{mix}S = -nR \left(x_{Ar} \ln x_{Ar} + x_{He} \ln x_{He} + x_{H_2} \ln x_{H_2} \right)$

$$= -(6 \text{ mol}) \left(8.314 \text{ J K}^{-1} \text{mol}^{-1} \right) \left(\frac{2}{6} \ln \frac{2}{6} + \frac{1}{6} \ln \frac{1}{6} + \frac{3}{6} \ln \frac{3}{6} \right) = 50.45 \text{ J K}^{-1}$$

5.12   Which of the following has a higher chemical potential? If neither, answer "same." **(a)** $H_2O(s)$ or $H_2O(l)$ at water's normal melting point, **(b)** $H_2O(s)$ at $-5°$ C and 1 bar or $H_2O(l)$ at $-5°$ C and 1 bar, **(c)** benzene at $25°$ C and 1 bar or benzene in a 0.1 M toluene solution in benzene at $25°$ C and 1 bar.

The less stable species of a pair has a higher chemical potential. When both substances in a pair are at equilibrium, then they have the same chemical potential.

(a) Same

(b) $H_2O(l)$ at $-5°$ C and 1 bar

(c) Benzene at $25°$ C and 1 bar. This is because $x_{benzene} < 0$ in the following relation:

$$\mu_{benzene} (l) = \mu_{benzene}^* (l) + RT \ln x_{benzene}$$

Therefore,

$$\mu_{benzene} (l) < \mu_{benzene}^* (l)$$

5.14   Write the phase equilibrium conditions for a liquid solution of methanol and water in equilibrium with its vapor.

At equilibrium, the chemical potentials for liquid and vapor for each component are equal. In this case, $\mu_{H_2O}(l) = \mu_{H_2O}(g)$ and $\mu_{CH_3OH}(l) = \mu_{CH_3OH}(g)$.

5.16   A miner working 900 ft below the surface had a soft drink beverage during the lunch break. To his surprise, the drink seemed very flat (that is, not much effervescence was observed upon removing the cap). Shortly after lunch, he took the elevator up to the surface. During the trip up, he felt a great urge to belch. Explain.

At 900 ft below sea level, the total pressure, and likewise the partial pressure of CO_2 is greater than it is at or slightly above sea level, where presumably the soft drink was bottled. Thus, the solubility of CO_2 is increased, resulting in a higher concentration in the solution. Upon returning to the surface, the partial pressure returns to normal (*i.e.* decreases), the solubility of CO_2 is reduced, and gaseous CO_2 begins to come out of solution in the miner's stomach, which leads to a natural urge to let this excess gas escape.

5.18   The solubility of N_2 in blood at 37° C and a partial pressure of 0.80 atm is 5.6×10^{-4} mol L^{-1}. A deep-sea diver breathes compressed air with a partial pressure of N_2 equal to 4.0 atm. Assume that the total volume of blood in the body is 5.0 L. Calculate the amount of N_2 gas released (in liters) when the diver returns to the surface of water, where the partial pressure of N_2 is 0.80 atm.

Assuming N_2 is an ideal gas, the volume of N_2 released can be readily calculated from the number of moles of N_2 released, which in turn is related to the number of moles of N_2 dissolved in blood when the partial pressures of N_2 are 4.0 atm and 0.80 atm, respectively.

According to Henry's law, the solubility of a substance is proportional to the applied pressure. Therefore,

$$\frac{\text{Solubility of } N_2 \text{ when } P_{N_2} \text{ is 4.0 atm}}{\text{Solubility of } N_2 \text{ when } P_{N_2} \text{ is 0.80 atm}} = \frac{4.0 \text{ atm}}{0.80 \text{ atm}} = 5.0$$

$$\text{Solubility of } N_2 \text{ when } P_{N_2} \text{ is 4.0 atm} = 5.0 \left(5.6 \times 10^{-4} \text{ mol L}^{-1} \right) = 2.80 \times 10^{-3} \text{ mol}$$

When $P_{N_2} = 0.80$ atm,

$$\text{Number of moles of } N_2 \text{ in blood} = \left(5.6 \times 10^{-4} \text{ mol L}^{-1} \right) (5.0 \text{ L}) = 2.80 \times 10^{-3} \text{ mol}$$

When $P_{N_2} = 4.0$ atm,

$$\text{Number of moles of } N_2 \text{ in blood} = \left(2.80 \times 10^{-3} \text{ mol L}^{-1} \right) (5.0 \text{ L}) = 1.40 \times 10^{-2} \text{ mol}$$

When the diver returns to the surface of water,

$$\text{Number of moles of } N_2 \text{ released} = 1.40 \times 10^{-2} \text{ mol} - 2.80 \times 10^{-3} \text{ mol} = 1.12 \times 10^{-2} \text{ mol}$$

and therefore,

$$\text{Volume of N}_2 \text{ released} = \frac{\left(1.12 \times 10^{-2} \text{ mol}\right) \left(0.08206 \text{ L atm K}^{-1} \text{ mol}^{-1}\right) (310 \text{ K})}{0.80 \text{ atm}} = 0.36 \text{ L}$$

5.20   Liquids A (bp = $T_A{}^\circ$) and B (bp = $T_B{}^\circ$) form an ideal solution. Predict the range of boiling points of solutions formed by mixing different amounts of A and B.

The range of boiling points will be between $T_A{}^\circ$ and $T_B{}^\circ$.

5.22   Two beakers, 1 and 2, containing 50 mL of 0.10 M urea and 50 mL of 0.20 M urea, respectively, are placed under a tightly sealed bell jar at 298 K. Calculate the mole fraction of urea in the solutions at equilibrium. Assume ideal behavior. (*Hint*: Use Raoult's law and note that at equilibrium, the mole fraction of urea is the same in both solutions.)

First find the number of moles of urea in each beaker.

$$\text{\# moles urea} = (0.050 \text{ L}) (0.10 \text{ } M) = 0.00500 \text{ mol}$$

and the second

$$\text{\# moles urea} = (0.050 \text{ L}) (0.20 \text{ } M) = 0.0100 \text{ mol}$$

Finding the mole fraction requires the number of moles of water in each beaker as well. Assume that the volume of each solution is equal to the volume of water, which is to say $V_{\text{H}_2\text{O}} = 50$ mL. Thus,

$$\text{\# moles water} = \frac{(50 \text{ mL}) \left(1 \text{ g mL}^{-1}\right)}{18.01 \text{ g mol}^{-1}}$$

$$= 2.78 \text{ mol}$$

The mole fraction of urea in each beaker is then

$$x_1 = \frac{0.00500 \text{ mol}}{0.00500 \text{ mol} + 2.78 \text{ mol}} = 1.80 \times 10^{-3}$$

$$x_2 = \frac{0.0100 \text{ mol}}{0.0100 \text{ mol} + 2.78 \text{ mol}} = 3.58 \times 10^{-3}$$

Equilibrium is attained by the transfer of water (via water vapor) from the less concentrated solution to the more concentrated one until the mole fractions of urea are equal. At this point, the mole fractions of water in each beaker are also equal and Raoult's law implies that the vapor pressures of the water over each beaker are the same. Thus, there is no more net transfer of solvent between beakers. Let y be the number of moles of water transferred to reach equilibrium.

$$\frac{0.00500 \text{ mol}}{0.00500 \text{ mol} + 2.78 \text{ mol} - y} = \frac{0.0100 \text{ mol}}{0.0100 \text{ mol} + 2.78 \text{ mol} + y}$$

$$y = 0.927 \text{ mol}$$

and the mole fraction of urea is

$$\frac{0.0100 \text{ mol}}{0.0100 \text{ mol} + 2.78 \text{ mol} + 0.927 \text{ mol}} = 2.7 \times 10^{-3}$$

This solution to the problem assumes that the volume of water left in the bell jar as vapor is negligible compared to the volumes of the solutions. It is interesting to note that at equilibrium 16.7 mL of water has been transferred from one beaker to the other.

5.24   Trees in cold climates may be subjected to temperatures as low as $-60°$ C. Estimate the concentration of an aqueous solution in the body of the tree that would remain unfrozen at this temperature. Is this a reasonable concentration? Comment on your result.

The freezing point of water is depressed by $60°$ C or 60 K. The concentration of solute giving rise to this depression is

$$m_2 = \frac{\Delta T}{K_f} = \frac{60 \text{ K}}{1.86 \text{ K mol}^{-1} \text{kg}} = 32 \text{ mol kg}^{-1}$$

This is a very high concentration. Indeed it would be impossible for any species with a molar mass greater than approximately 31 g mol^{-1} to be present in a solution at this concentration. (In such a case, 32 moles of the solute would have a mass greater than 1 kg.) Indeed, some of the species present are certainly electrolytes, which would lower the concentration required to attain such a freezing point depression. Additionally, the core of the tree is insulated by the bark, so that the full 60 K depression may not be required. Finally, living systems that are adapted to such harsh conditions may produce proteins that inhibit the formation of ice crystals. Such "antifreeze" proteins are well-studied in fish, whether cold-adapted trees possess similar proteins is an open question.

5.26   Provide a molecular interpretation for the positive and negative deviations in the boiling-point curves.

If the intermolecular forces between unlike molecules are greater (more attractive) than those between like molecules, the solution will show a positive deviation in the boiling point curve. On the other hand, if the intermolecular forces are weaker between unlike molecules than between like molecules there will be a negative deviation.

5.28   A common antifreeze for car radiators is ethylene glycol, $CH_2(OH)CH_2(OH)$. How many milliliters of this substance would you add to 6.5 L of water in the radiator if the coldest day in winter is $-20°$ C? Would you keep this substance in the radiator in the summer to prevent the water from boiling? (The density and boiling point of ethylene glycol are 1.11 g cm^{-3} and 470 K, respectively.)

The molality of ethylene glycol that can depress the freezing point by $20°$ C or 20 K is

$$m_2 = \frac{\Delta T}{K_f} = \frac{20 \text{ K}}{1.86 \text{ K mol}^{-1} \text{kg}} = 10.75 \text{ mol kg}^{-1}$$

Assuming the density of water is 1 kg L^{-1}, the mass of water in the radiator is 6.5 kg, which, together with the molality of ethylene glycol, gives the number of moles of ethylene glycol.

$$\text{Number of moles of ethylene glycol} = \left(10.75 \text{ mol kg}^{-1}\right)(6.5 \text{ kg}) = 69.88 \text{ mol}$$

which corresponds to a volume of

$$V = \frac{(69.88 \text{ mol}) \left(62.07 \text{ g mol}^{-1}\right)}{1.11 \text{ g cm}^{-3}} = 3.91 \times 10^3 \text{ cm}^3 = 3.91 \times 10^3 \text{ mL}$$

Since it has a higher boiling point, ethylene glycol can also elevate the boiling point of water. With a 10.75 m solution, the boiling point of water will be elevated by

$$\Delta T = K_b m_2 = \left(0.51 \text{ K mol}^{-1} \text{kg}\right) \left(10.75 \text{ mol kg}^{-1}\right) = 5.5 \text{ K}$$

Therefore, keeping ethylene glycol in the radiator in the summer will increase the boiling point by 5.5° C.

5.30   The tallest trees known are the redwoods in California. Assuming the height of a redwood to be 105 m (about 350 ft), estimate the osmotic pressure required to push water up from the roots to the treetop.

Assume that the density of water is 1×10^3 kg m^{-3}.

$$\pi = h\rho g \qquad \text{(See Example 5.6 in text.)}$$
$$= (105 \text{ m}) \left(1 \times 10^3 \text{ kg m}^{-3}\right) \left(9.81 \text{ m s}^{-2}\right)$$
$$= \left(1.030 \times 10^6 \text{ N m}^{-2}\right) \left(\frac{1 \text{ atm}}{1.013 \times 10^5 \text{ N m}^{-2}}\right) = 10.2 \text{ atm}$$

5.32   Fish breathe the dissolved air in water through their gills. Assuming the partial pressures of oxygen and nitrogen in air to be 0.20 atm and 0.80 atm, respectively, calculate the mole fractions of oxygen and nitrogen in water at 298 K. Comment on your results.

The mole fractions of oxygen and nitrogen can be calculated using Henry's law (See Table 5.1).

$$x_{O_2} = \frac{P_{O_2}}{K_{O_2}} = \left(\frac{0.20 \text{ atm}}{3.27 \times 10^7 \text{ torr}}\right) \left(\frac{760 \text{ torr}}{1 \text{ atm}}\right) = 4.6 \times 10^{-6}$$

$$x_{N_2} = \frac{P_{N_2}}{K_{N_2}} = \left(\frac{0.80 \text{ atm}}{6.80 \times 10^7 \text{ torr}}\right) \left(\frac{760 \text{ torr}}{1 \text{ atm}}\right) = 8.9 \times 10^{-6}$$

In water, the concentration of N_2 is just twice that of O_2. In the atmosphere, $\frac{1}{5}$ of the gas is O_2, but in water $\frac{1}{3}$ of the dissolved gas is O_2.

5.34   Lysozyme extracted from chicken egg white has a molar mass of 13,930 g mol^{-1}. Exactly 0.1 g of this protein is dissolved in 50 g of water at 298 K. Calculate the vapor pressure lowering, the depression in freezing point, the elevation of boiling point, and the osmotic pressure of this solution. The vapor pressure of pure water at 298 K is 23.76 mmHg.

First calculate the mole fraction, molality, and molarity of lysozyme.

$$\text{Number of moles of lysozyme} = \frac{0.1 \text{ g}}{13930 \text{ g mol}^{-1}} = 7.17875 \times 10^{-6} \text{ mol}$$

$$\text{Number of moles of water} = \frac{50 \text{ g}}{18.02 \text{ g mol}^{-1}} = 2.7747 \text{ mol}$$

$$x_{\text{lysozyme}} = \frac{7.17875 \times 10^{-6} \text{ mol}}{7.17875 \times 10^{-6} \text{ mol} + 2.7747 \text{ mol}} = 2.5872 \times 10^{-6}$$

$$m = \frac{7.17875 \times 10^{-6} \text{ mol}}{50 \times 10^{-3} \text{ kg}} = 1.43575 \times 10^{-4} \text{ mol kg}^{-1}$$

For a dilute aqueous solution, the molality and molarity are numerically the same (see Problem 5.6). Therefore, $M = 1.43575 \times 10^{-4}$ mol L^{-1}.

Vapor pressure lowering:

$$\Delta P = x_{\text{lysozyme}} P_{\text{H}_2\text{O}}^* = \left(2.5872 \times 10^{-6}\right)(23.76 \text{ mmHg}) = 6.147 \times 10^{-5} \text{ mmHg}$$

Depression in freezing point:

$$\Delta T_f = K_f m = \left(1.86 \text{ K m}^{-1}\right)\left(1.43575 \times 10^{-4} \text{ m}\right) = 2.67 \times 10^{-4} \text{ K}$$

Elevation of boiling point:

$$\Delta T_b = K_b m = \left(0.51 \text{ K m}^{-1}\right)\left(1.43575 \times 10^{-4} \text{ m}\right) = 7.3 \times 10^{-5} \text{ K}$$

Osmotic pressure:

$$\pi = MRT = \left(1.43575 \times 10^{-4} \text{ M}\right)\left(0.08206 \text{ L atm K}^{-1} \text{ mol}^{-1}\right)(298 \text{ K})$$

$$= 3.51 \times 10^{-3} \text{ atm} \left(\frac{760 \text{ torr}}{1 \text{ atm}}\right) = 2.67 \text{ torr}$$

Note that the only property that is readily measurable is the osmotic pressure.

5.36   A compound weighing 0.458 g is dissolved in 30.0 g of acetic acid. The freezing point of the solution is found to be 1.50 K below that of the pure solvent. Calculate the molar mass of the compound.

The molality of the solution is related to the depression in freezing point.

$$m = \frac{\Delta T}{K_f} = \frac{1.50 \text{ K}}{3.90 \text{ K mol}^{-1} \text{ kg}} = 0.3846 \text{ mol kg}^{-1}$$

The number of moles of the compound is

$$n = \left(0.3846 \text{ mol kg}^{-1}\right)\left(30.0 \times 10^{-3} \text{ kg}\right) = 0.01154 \text{ mol}$$

Therefore,

$$\text{Molar mass of the compound} = \frac{0.458 \text{ g}}{0.01154 \text{ mol}} = 39.7 \text{ g mol}^{-1}$$

5.38 A forensic chemist is given a white powder for analysis. She dissolves 0.50 g of the substance in 8.0 g of benzene. The solution freezes at 3.9° C. Can the chemist conclude that the compound is cocaine ($C_{17}H_{21}NO_4$)? What assumptions are made in the analysis? The freezing point of benzene is 5.5° C.

The depression in freezing point of benzene (5.5° C − 3.9 °C = 1.6° C = 1.6 K) furnishes the molar mass of the white powder, which is then compared with the molar mass of cocaine.

The molality of the compound is

$$m = \frac{\Delta T}{K_f} = \frac{1.6 \text{ K}}{5.12 \text{ K mol}^{-1} \text{ kg}} = 0.313 \text{ mol kg}^{-1}$$

The number of moles of the compound is

$$n = \left(0.313 \text{ mol kg}^{-1}\right)\left(8.0 \times 10^{-3} \text{ kg}\right) = 2.50 \times 10^{-3} \text{ mol}$$

Therefore, the molar mass of the substance is

$$\mathcal{M} = \frac{0.50 \text{ g}}{2.50 \times 10^{-3} \text{ mol}} = 2.0 \times 10^2 \text{ g mol}^{-1}$$

The molar mass of cocaine is 303.35 g mol^{-1}. Thus, the compound is not likely to be cocaine. The assumptions implicit in this analysis are that the compound is pure, that it is monomeric and does not either associate or dissociate in benzene, and that it is not an electrolyte.

5.40 A nonvolatile organic compound, Z, was used to make up two solutions. Solution A contains 5.00 g of Z dissolved in 100 g of water, and solution B contains 2.31 g of Z dissolved in 100 g of benzene. Solution A has a vapor pressure of 754.5 mmHg at the normal boiling point of water, and solution B has the same vapor pressure at the normal boiling point of benzene. Calculate the molar mass of Z in solutions A and B, and account for the difference.

Since Z causes the same amount of vapor pressure lowering in Solutions A and B, the mole fraction of Z, x_Z, must be the same in both solutions.

$$x_Z = \frac{\Delta P}{P^*} = \frac{760 \text{ mmHg} - 754.5 \text{ mmHg}}{760 \text{ mmHg}} = 7.24 \times 10^{-3}$$

The molar mass of Z in each solution can then be calculated from the number of moles of solvent, the mole fraction of Z, and the mass of Z.

Solution A

$$n_{H_2O} = \frac{100 \text{ g}}{18.02 \text{ g mol}^{-1}} = 5.549 \text{ mol}$$

$$x_Z = 7.24 \times 10^{-3} = \frac{n_Z}{n_Z + n_{H_2O}}$$

$$n_Z = 7.24 \times 10^{-3} \left(n_Z + n_{H_2O}\right)$$

$$0.99276 n_Z = 7.24 \times 10^{-3} n_{H_2O} = 0.0402$$

$$n_Z = 4.05 \times 10^{-2} \text{ mol}$$

$$\text{Molar mass of Z} = \frac{5.00 \text{ g}}{4.05 \times 10^{-2} \text{ mol}} = 1.2 \times 10^2 \text{ g mol}^{-1}$$

Solution B

$$n_{\text{benzene}} = \frac{100 \text{ g}}{78.11 \text{ g mol}^{-1}} = 1.280 \text{ mol}$$

$$x_Z = 7.24 \times 10^{-3} = \frac{n_Z}{n_Z + n_{\text{benzene}}}$$

$$n_Z = 7.24 \times 10^{-3} \left(n_Z + n_{\text{benzene}}\right)$$

$$0.99276 n_Z = 7.24 \times 10^{-3} n_{\text{benzene}} = 0.00927$$

$$n_Z = 9.34 \times 10^{-3} \text{ mol}$$

$$\text{Molar mass of Z} = \frac{2.31 \text{ g}}{9.34 \times 10^{-3} \text{ mol}} = 2.5 \times 10^2 \text{ g mol}^{-1}$$

The molar mass of Z in benzene is about twice that in water. This suggests some sort of dimerization is occurring in a nonpolar solvent such as benzene.

5.42 At 85° C, the vapor pressure of A is 566 torr and that of B is 250 torr. Calculate the composition of a mixture of A and B that boils at 85° C when the pressure is 0.60 atm. Also, calculate the composition of the vapor mixture. Assume ideal behavior.

The total vapor pressure is the same as the external pressure, 0.60 atm, when the mixture boils. The total vapor pressure is the sum of the vapor pressures of A and B from the mixture, which depend on the mole fractions of A and B in the mixture.

$$P_{\text{total}} = (0.60 \text{ atm}) \left(\frac{760 \text{ torr}}{1 \text{ atm}}\right) = P_A + P_B = x_A P_A^* + x_B P_B^*$$

$$456 \text{ torr} = x_A (566 \text{ torr}) + x_B (250 \text{ torr})$$

Since $x_B = 1 - x_A$, the above equation depends on only one variable, x_A.

$$456 \text{ torr} = x_A \, (566 \text{ torr}) + \left(1 - x_A\right) (250 \text{ torr})$$

$$(316 \text{ torr}) \, x_A = 206 \text{ torr}$$

$$x_A = 0.652 = 0.65$$

Therefore,

$$x_B = 1 - x_A = 0.348 = 0.35$$

The mole fractions of A and B are 0.65 and 0.35, respectively, in the mixture. The composition in the vapor can be evaluated from the vapor pressures of A and B.

$$P_A = x_A P_A^* = (0.652) \, (566 \text{ torr}) = 369 \text{ torr}$$

$$P_B = P_{\text{total}} - P_A = 456 \text{ torr} - 369 \text{ torr} = 87 \text{ torr}$$

$$\frac{P_A}{P_B} = \frac{369 \text{ torr}}{87 \text{ torr}} = \frac{x_A^v P_{\text{total}}}{x_B^v P_{\text{total}}} = \frac{x_A^v}{x_B^v}$$

$$\frac{x_A^v}{x_B^v} = 4.24$$

$$x_A^v = 4.24 x_B^v = 4.24 \left(1 - x_A^v\right)$$

$$5.24 x_A^v = 4.24$$

$$x_A^v = 0.81$$

$$x_B^v = 0.19$$

The vapor of the mixture is richer in A, the more volatile component.

5.44   Liquids A and B form an ideal solution at a certain temperature. The vapor pressures of pure A and B are 450 torr and 732 torr, respectively, at this temperature. (**a**) A sample of the solution's vapor is condensed. Given that the original solution contains 3.3 moles of A and 8.7 moles of B, calculate the composition of the condensate in mole fractions. (**b**) Suggest a method for measuring the partial pressures of A and B at equilibrium.

(**a**) First calculate the vapor pressures of A and B using Raoult's law.

$$x_A = \frac{3.3 \text{ mol}}{3.3 \text{ mol} + 8.7 \text{ mol}} = 0.275$$

$$x_B = 1 - 0.275 = 0.725$$

$$P_A = x_A P_A^* = (0.275) \, (450 \text{ torr}) = 124 \text{ torr}$$

$$P_B = x_B P_B^* = (0.725) \, (732 \text{ torr}) = 531 \text{ torr}$$

The composition of the condensed liquid is the same as the composition of the vapor before condensation, the latter can be calculated from P_A and P_B.

$$\frac{P_A}{P_B} = \frac{124 \text{ torr}}{531 \text{ torr}} = \frac{x_A^v P_{total}}{x_B^v P_{total}} = \frac{x_A^v}{x_B^v}$$

$$\frac{x_A^v}{x_B^v} = 0.234$$

$$x_A^v = 0.234 x_B^v = 0.234 \left(1 - x_A^v\right)$$

$$1.234 x_A^v = 0.234$$

$$x_A^v = 0.19$$

$$x_B^v = 0.81$$

(b) Using Dalton's Law, knowledge of the mole fractions of A and B in the vapor phase will allow the determination of their partial pressures via $P_i = x_i P_{total}$. Thus, mass spectrometric analysis of a sample of the vapor in equilibrium with the solution to determine the mole fraction of each plus a measurement of the total pressure will provide the needed information.

5.46 Calculate the molal boiling-point elevation constant (K_b) for water. The molar enthalpy of vaporization of water is 40.79 kJ mol^{-1} at 100° C.

$$K_b = \frac{RT_0^2 \mathcal{M}_1}{\Delta_{vap}\overline{H}}$$

$$= \frac{\left(8.314 \text{ J K}^{-1} \text{ mol}^{-1}\right)\left(373.15 \text{ K}\right)^2 \left(18.02 \times 10^{-3} \text{ kg mol}^{-1}\right)}{40.79 \times 10^3 \text{ J mol}^{-1}}$$

$$= 0.5114 \text{ K kg mol}^{-1} = 0.5114 \text{ K m}^{-1}$$

5.48 The following data give the pressures for carbon disulfide–acetone solutions at 35.2° C. Calculate the activity coefficients of both components based on deviations from Raoult's law and Henry's law. (*Hint*: First determine Henry's law constants graphically.)

x_{CS_2}	0	0.20	0.45	0.67	0.83	1.00
P_{CS_2}/torr	0	272	390	438	465	512
$P_{C_3H_6O}$/torr	344	291	250	217	180	0

From a graph of the vapor pressure data, Henry's law constants may be obtained from the limiting slopes of the curves as they approach their respective infinite dilution (mole fraction equals zero) limits. (Note that the graph pictured is scaled to show the $x = 1$ intercept of each Henry's law line. In making the actual determination of the limiting slopes it is better to use an expanded scale that shows the limiting behavior in detail.) The Henry's law constants obtained from these data are $K_{CS_2} = 1570$ torr and $K_{C_3H_6O} = 1250$ torr. The activity coefficient of the ith component accounts for deviations from Henry's law as

$$P_i = K_i \gamma_i x_i$$

or

$$\gamma_i = \frac{P_i}{K_i x_i}$$

The necessary data to find the Henry's law activity coefficients, $\gamma_i(H)$ are in the table. (Note that $x_{C_3H_6O} = 1 - x_{CS_2}$.)

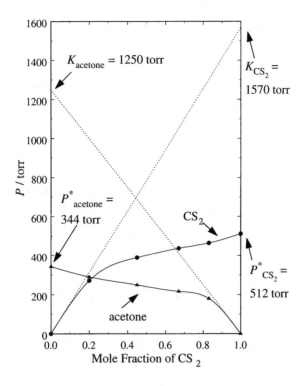

x_{CS_2}	0	0.20	0.45	0.67	0.83	1.00
P_{CS_2}/torr	0	272	390	438	465	512
$P_{C_3H_6O}$/torr	344	291	250	217	180	0
$\gamma_{CS_2}(H)$	1.00	0.87	0.55	0.42	0.36	0.33
$\gamma_{C_3H_6O}(H)$	0.28	0.29	0.36	0.53	0.85	1.00

The Raoult's law activity is found through deviations from Raoult's law via

$$P_i = \gamma_i x_i P_i^*$$

or

$$\gamma_i = \frac{P_i}{x_i P_i^*}$$

The tabulated data are used once again, noting that the vapor pressures of the pure components are included, namely $P_{CS_2}^* = 512$ torr and $P_{C_3H_6O}^* = 344$ torr. The Raoult's law activity is denoted $\gamma_i(R)$, and is not defined at zero mole fraction.

x_{CS_2}	0	0.20	0.45	0.67	0.83	1.00
P_{CS_2}/torr	0	272	390	438	465	512
$P_{C_3H_6O}$/torr	344	291	250	217	180	0
$\gamma_{CS_2}(R)$	—	2.66	1.69	1.28	1.09	1.00
$\gamma_{C_3H_6O}(R)$	1.00	1.06	1.32	1.91	3.08	—

The activities determined from the two laws differ because the standard state is chosen differently in the two cases.

5.50 A certain dilute solution has an osmotic pressure of 12.2 atm at 20° C. Calculate the difference between the chemical potential of the solvent in the solution and that of pure water. Assume that the density is the same as that of water. (*Hint*: Express the chemical potential in terms of mole fraction, x_1, and rewrite the osmotic pressure equation as $\pi V = n_2 RT$, where n_2 is the number of moles of the solute and $V = 1$ L.)

Let the chemical potential of pure water be $\mu_1^* (l)$ and the chemical potential of the solvent be $\mu_1 (l)$. These two chemical potentials are related by

$$\mu_1 (l) = \mu_1^* (l) + RT \ln x_1$$

The mole fraction of water, x_1, is then calculated from the number of moles of solute, n_2, and the number of moles of water, n_1, in 1 L of solution. The number of moles of solute is related to the osmotic pressure of the solution.

$$\pi = MRT = \frac{n_2}{V} RT$$

$$n_2 = \frac{\pi V}{RT} = \frac{(12.2 \text{ atm}) (1 \text{ L})}{\left(0.08206 \text{ L atm K}^{-1} \text{ mol}^{-1}\right) (293.2 \text{ K})} = 0.5071 \text{ mol}$$

Since the solution is dilute, the volume of water is assumed to be the same as the volume of the solution and the density of water is assumed to be 1 g cm^{-3}. Then

$$n_1 = \frac{\left(1000 \text{ cm}^3\right) \left(1 \text{ g cm}^{-3}\right)}{18.02 \text{ g mol}^{-1}} = 55.494 \text{ mol}$$

The mole fraction of water is

$$x_1 = \frac{n_1}{n_1 + n_2} = \frac{55.494 \text{ mol}}{55.494 \text{ mol} + 0.5071 \text{ mol}} = 0.9909$$

Substituting x_1 into the chemical potential expression above,

$$\mu_1 (l) = \mu_1^* (l) + \left(8.314 \text{ J K}^{-1} \text{ mol}^{-1}\right) (293.2 \text{ K}) \ln (0.9909) = \mu_1^* (l) - 22.3 \text{ J mol}^{-1}$$

The chemical potential of the solution is lower than that of water by 22.3 J mol^{-1}.

5.52 Consider a binary liquid mixture A and B, where A is volatile and B is nonvolatile. The composition of the solution in terms of mole fraction is $x_A = 0.045$ and $x_B = 0.955$. The vapor pressure of A from the mixture is 5.60 mmHg, and that of pure A is 196.4 mmHg at the same temperature. Calculate the activity coefficient of A at this concentration.

The activity of A is

$$a_A = \frac{P_A}{P_A^*} = \frac{5.60 \text{ mmHg}}{196.4 \text{ mmHg}} = 2.851 \times 10^{-2}$$

The activity coefficient of A is

$$\gamma_A = \frac{a_A}{x_A} = \frac{2.851 \times 10^{-2}}{0.045} = 0.63$$

5.54 Calculate the ionic strength and the mean activity coefficient for the following solutions at 298 K: **(a)** 0.10 m NaCl, **(b)** 0.010 m MgCl$_2$, and **(c)** 0.10 m K$_4$Fe(CN)$_6$.

The ionic strength can be obtained from the equation

$$I = \frac{1}{2} \sum_i m_i z_i^2$$

and subsequently the mean activity from the Debye-Hückel limiting law

$$\log \gamma_\pm = -0.509 \left| z_+ z_- \right| \sqrt{I}$$

(a) 0.10 m NaCl: $z_+ = 1$, $z_- = -1$, $m_+ = 0.10\ m$, $m_- = 0.10\ m$

$$I = \frac{1}{2} \left[(0.10\ m)\,(1)^2 + (0.10\ m)\,(-1)^2 \right] = 0.10\ m$$

$$\log \gamma_\pm = -0.509\,|(1)\,(-1)|\,\sqrt{0.10} = -0.161$$

$$\gamma_\pm = 0.69$$

(b) 0.010 m MgCl$_2$: $z_+ = 2$, $z_- = -1$, $m_+ = 0.010\ m$, $m_- = 0.020\ m$

$$I = \frac{1}{2} \left[(0.010\ m)\,(2)^2 + (0.020\ m)\,(-1)^2 \right] = 0.030\ m$$

$$\log \gamma_\pm = -0.509\,|(2)\,(-1)|\,\sqrt{0.030} = -0.176$$

$$\gamma_\pm = 0.67$$

(c) 0.10 m K$_4$Fe(CN)$_6$: $z_+ = 1$, $z_- = -4$, $m_+ = 0.40\ m$, $m_- = 0.10\ m$

$$I = \frac{1}{2} \left[(0.40\ m)\,(1)^2 + (0.10\ m)\,(-4)^2 \right] = 1.0\ m$$

$$\log \gamma_\pm = -0.509\,|(1)\,(-4)|\,\sqrt{1.0} = -2.04$$

$$\gamma_\pm = 9.1 \times 10^{-3}$$

5.56 A 0.20 m Mg(NO$_3$)$_2$ solution has a mean ionic activity coefficient of 0.13 at 25° C. Calculate the mean molality, the mean ionic activity, and the activity of the compound.

For the Mg(NO$_3$)$_2$ solution, $\nu_+ = 1$, $\nu_- = 2$, $\nu = 3$, $m_+ = 0.20\ m$, $m_- = 0.40\ m$. The mean molality is

$$m_\pm = \left[m_+^{\nu_+} m_-^{\nu_-} \right]^{1/\nu} = \left[(0.20\ m)\,(0.40\ m)^2 \right]^{1/3} = 0.317\ m = 0.32\ m$$

The mean ionic activity is

$$a_\pm = \gamma_\pm m_\pm = (0.13)\,(0.317) = 0.0412 = 0.041$$

The activity is

$$a = a_\pm^\nu = (0.0412)^3 = 7.0 \times 10^{-5}$$

5.58 In theory, the size of the ionic atmosphere is $1/\kappa$, called the Debye radius, and κ is given by

$$\kappa = \left(\frac{e^2 N_A}{\epsilon_0 \epsilon k_B T} \right)^{1/2} \sqrt{I}$$

where e is the electronic charge, N_A Avogadro's constant, ϵ_0 the permittivity of vacuum ($8.854 \times 10^{-12}\ C^2\,N^{-1}\,m^{-2}$), ϵ the dielectric constant of the solvent, k_B the Boltzmann constant, T the absolute temperature, and I the ionic strength (see the physical chemistry texts listed in Chapter 1). Calculate the Debye radius in a $0.010\ m$ aqueous Na_2SO_4 solution at $25°$ C.

The ionic strength of a $0.010\ m$ Na_2SO_4 solution is

$$I = \frac{1}{2} \sum_i m_i z_i^2$$

$$= \frac{1}{2} \left[(0.020\ m)\,(1)^2 + (0.010\ m)\,(-2)^2 \right]$$

$$= 0.030\ m$$

Since this is a dilute solution, the ionic strength in $mol\,L^{-1}$ can be taken as numerically equal to that in $mol\,kg^{-1}$, or $0.030\ M$.

For water, $\epsilon = 78.54$. Thus,

$$\kappa = \left(\frac{e^2 N_A}{\epsilon_0 \epsilon k_B T} \right)^{1/2} \sqrt{I}$$

$$= \left[\frac{\left(1.602 \times 10^{-19}\ C\right)^2 \left(6.022 \times 10^{23}\ mol^{-1}\right)}{\left(8.854 \times 10^{-12}\ C^2\,N^{-1}\,m^{-2}\right) (78.54) \left(1.381 \times 10^{-23}\ J\,K^{-1}\right) (298\ K)} \right]^{1/2}$$

$$\times \sqrt{\left(0.030\ mol\,L^{-1}\right) \left(\frac{1000\ L}{1\ m^3} \right)}$$

$$= 4.025 \times 10^8\ m^{-1}$$

$$\frac{1}{\kappa} = 2.48 \times 10^{-9}\ m = 24.8\ \mathring{A}$$

5.60 Calculate the ionic strength of a $0.0020\ m$ aqueous solution of $MgCl_2$ at 298 K. Use the Debye–Hückel limiting law to estimate **(a)** the activity coefficients of the Mg^{2+} and Cl^- ions in this solution and **(b)** the mean ionic activity coefficients of these ions.

The ionic strength of the solution is

$$I = \frac{1}{2} \sum_i m_i z_i^2 = \frac{1}{2} \left[(0.0020 \, m) \, (2)^2 + (0.0040 \, m) \, (-1)^2 \right] = 0.0060 \, m$$

(a) The activity coefficients of Mg^{2+} and Cl^- can be evaluated using

$$\log \gamma_i = -0.509 z_i^2 \sqrt{I}$$

For Mg^{2+},

$$\log \gamma_+ = -0.509 \, (2)^2 \, \sqrt{0.0060} = -0.158$$

$$\gamma_+ = 0.695 = 0.70$$

For Cl^-,

$$\log \gamma_- = -0.509 \, (-1)^2 \, \sqrt{0.0060} = -3.94 \times 10^{-2}$$

$$\gamma_- = 0.913 = 0.91$$

(b) The mean ionic activity coefficient is

$$\gamma_\pm = \left(\gamma_+^{\nu_+} \gamma_-^{\nu_-} \right)^{1/\nu} = \left[(0.695) \, (0.913)^2 \right]^{1/3} = 0.83$$

5.62   **(a)** Which of the following expressions is incorrect as a representation of the partial molar volume of component A in a two-component solution? Why? How would you correct it?

$$\left(\frac{\partial V_m}{\partial n_A} \right)_{T,P,n_B} \qquad\qquad \left(\frac{\partial V_m}{\partial x_A} \right)_{T,P,x_B}$$

(b) Given that the molar volume of this mixture (V_m) is given by

$$V_m = 0.34 + 3.6 x_A x_B + 0.4 x_B \left(1 - x_A \right) \, \text{L mol}^{-1}$$

derive an expression for the partial molar volume for A at $x_A = 0.20$.

(a) The second expression is incorrect, since it is impossible to vary x_A while holding x_B constant. The correct expression is $\left(\frac{\partial V_m}{\partial x_A} \right)_{T,P}$

(b) Write V_m in terms of x_A.

$$V_m = \left[0.34 + 3.6 x_A \left(1 - x_A \right) + 0.4 \left(1 - x_A \right) \left(1 - x_A \right) \right] \, \text{L mol}^{-1}$$

$$= \left[0.34 + 3.6 x_A - 3.6 x_A^2 + 0.4 - 0.8 x_A + 0.4 x_A^2 \right] \, \text{L mol}^{-1}$$

$$= \left(0.38 + 2.8 x_A - 3.2 x_A^2 \right) \, \text{L mol}^{-1}$$

Differentiate V_m with respect to x_A.

$$\left(\frac{\partial V_m}{\partial x_A} \right)_{P,T} = \left(2.8 - 6.4 x_A \right) \, \text{L mol}^{-1}$$

At $x_A = 0.20$,

$$\left(\frac{\partial V_m}{\partial x_A}\right)_{P,T} = [2.8 - 6.4\,(0.20)] \text{ L mol}^{-1} = 1.5 \text{ L mol}^{-1}$$

5.64 The osmotic pressure of poly(methyl methacrylate) in toluene has been measured at a series of concentrations at 298 K. Determine graphically the molar mass of the polymer.

π/ atm	8.40×10^{-4}	1.72×10^{-3}	2.52×10^{-3}	3.23×10^{-3}	7.75×10^{-3}
c/g·L^{-1}	8.10	12.31	15.00	18.17	28.05

In the limit of a dilute solution, where all virial coefficients except the second may be ignored, the osmotic pressure π is given by

$$\frac{\pi}{c} = \frac{RT}{\mathcal{M}}\,(1 + Bc)$$

Thus, a graph of π/c vs. c will extrapolate to a y intercept of $\frac{RT}{\mathcal{M}}$, where $\mathcal{M}$ is the molar mass of the solute.

The data for the graph is given in the table below.

(π/c)/ atm·L·g^{-1}	1.037×10^{-4}	1.397×10^{-4}	1.680×10^{-4}	1.778×10^{-4}	2.763×10^{-4}
c/g·L^{-1}	8.10	12.31	15.00	18.17	28.05

As seen in the graph, the extrapolation to zero concentration gives an intercept of 3.4×10^{-5} atm L g^{-1}.

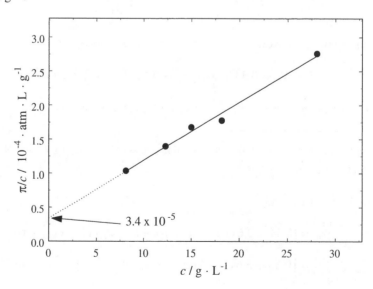

Thus,

$$3.4 \times 10^{-5} \text{atm L g}^{-1} = \frac{RT}{\mathcal{M}}$$

$$= \frac{\left(0.08206 \text{ L atm K}^{-1} \text{ mol}^{-1}\right)(298 \text{ K})}{\mathcal{M}}$$

$$\mathcal{M} = 7.2 \times 10^5 \text{ g mol}^{-1}$$

5.66 Suppose 2.6 moles of He at 0.80 atm and 25° C are mixed with 4.1 moles of Ne at 2.7 atm and 25° C. Calculate the Gibbs energy change for the process. Assume ideal behavior.

The expression $\Delta_{mix}G = nRT\left(x_1 \ln x_1 + x_2 \ln x_2\right)$ is valid only for mixing two gases originally at the same pressure and is not directly applicable here. G is a state function, however, so consider a path in which the two gases are first individually brought to the final pressure of the mixture and then mixed. The overall ΔG is the sum of the ΔG's for the steps. The process is assumed to take place with the gases originally in two containers connected by a valve which is opened. The final volume of the mixture is then the sum of the original volumes. Since $V_{He} = \frac{n_{He}RT}{P_{He}}$, with an similar expression for Ne,

$$P_f = \frac{n_{total}RT}{V_f}$$

$$= \frac{\left(n_{He} + n_{Ne}\right)RT}{\frac{n_{He}RT}{P_{He}} + \frac{n_{Ne}RT}{P_{Ne}}}$$

$$= \frac{n_{He} + n_{Ne}}{\frac{n_{He}}{P_{He}} + \frac{n_{Ne}}{P_{Ne}}}$$

$$= \frac{2.6\ \text{mol} + 4.1\ \text{mol}}{\frac{2.6\ \text{mol}}{0.80\ \text{atm}} + \frac{4.1\ \text{mol}}{2.7\ \text{atm}}}$$

$$= 1.41\ \text{atm}$$

For the compression (expansion) of an ideal gas, $\Delta G = nRT \ln \frac{P_f}{P_i}$, so that

$$\Delta G_{He} = (2.6\ \text{mol})\left(8.314\ \text{J K}^{-1}\text{mol}^{-1}\right)(298\ \text{K})\ln\frac{1.41\ \text{atm}}{0.80\ \text{atm}} = 3.65 \times 10^3\ \text{J}$$

$$\Delta G_{Ne} = (4.1\ \text{mol})\left(8.314\ \text{J K}^{-1}\text{mol}^{-1}\right)(298\ \text{K})\ln\frac{1.41\ \text{atm}}{2.7\ \text{atm}} = -6.60 \times 10^3\ \text{J}$$

$$\Delta_{mix}G = n_{total}RT\left(x_{He} \ln x_{He} + x_{Ne} \ln x_{Ne}\right)$$

$$= (2.6\ \text{mol} + 4.1\ \text{mol})\left(8.314\ \text{J K}^{-1}\text{mol}^{-1}\right)(298\ \text{K})$$

$$\times \left[\frac{2.6\ \text{mol}}{2.6\ \text{mol} + 4.1\text{mol}}\ln\left(\frac{2.6\ \text{mol}}{2.6\ \text{mol} + 4.1\text{mol}}\right)\right.$$

$$\left. + \frac{4.1\ \text{mol}}{2.6\ \text{mol} + 4.1\text{mol}}\ln\left(\frac{4.1\ \text{mol}}{2.6\ \text{mol} + 4.1\text{mol}}\right)\right]$$

$$= -1.11 \times 10^4\ \text{J}$$

$$\Delta G_{total} = \Delta G_{He} + \Delta G_{Ne} + \Delta_{mix}G = -1.4 \times 10^4\ \text{J} = -14\ \text{kJ}$$

5.68 From the following data, calculate the heat of solution for KI:

	NaCl	NaI	KCl	KI
Lattice energy/kJ·mol^{-1}	787	700	716	643
Heat of solution/kJ·mol^{-1}	3.8	−5.1	17.1	?

Begin by using $\Delta_{soln}H = U_0 + \Delta_{hydr}H$, where U_0 is the lattice energy.

$$Na^+(g) + Cl^-(g) \longrightarrow Na^+(aq) + Cl^-(aq)$$

$$\Delta_{hydr}H = 3.8 \text{ kJ mol}^{-1} - 787 \text{ kJ mol}^{-1} = -783.2 \text{ kJ mol}^{-1} \qquad (5.68.1)$$

$$Na^+(g) + I^-(g) \longrightarrow Na^+(aq) + I^-(aq)$$

$$\Delta_{hydr}H = -5.1 \text{ kJ mol}^{-1} - 700 \text{ kJ mol}^{-1} = -705.1 \text{ kJ mol}^{-1} \qquad (5.68.2)$$

$$K^+(g) + Cl^-(g) \longrightarrow K^+(aq) + Cl^-(aq)$$

$$\Delta_{hydr}H = 17.1 \text{ kJ mol}^{-1} - 716 \text{ kJ mol}^{-1} = -698.9 \text{ kJ mol}^{-1} \qquad (5.68.3)$$

Taking Equations 5.68.2 plus 5.68.3 minus 5.68.1 results in

$$K^+(g) + I^-(g) \longrightarrow K^+(aq) + I^-(aq)$$

$$\Delta_{hydr}H = -705.1 \text{ kJ mol}^{-1} - 698.9 \text{ kJ mol}^{-1} + 783.2 \text{ kJ mol}^{-1} = -620.8 \text{ kJ mol}^{-1}$$

Combine this last result with the given value of the lattice energy to arrive at the desired heat of solution.

$$\Delta_{soln}H = U_0 + \Delta_{hydr}H$$

$$= 643 \text{ kJ mol}^{-1} - 620.8 \text{ kJ mol}^{-1}$$

$$= 22.2 \text{ kJ mol}^{-1}$$

5.70 In this chapter (see Figures 5.12, 5.17, and 5.19), we extrapolated concentration-dependent values to zero solute concentration. Explain what these extrapolated values mean physically and why they differ from the value obtained for the pure solvent.

Extrapolating a concentration-dependent quantity to zero solute concentration corresponds to the value the quantity would take in the absence of solute-solute interactions. This infinitely dilute solution is not the same as pure solvent, since the solute molecules are still there. The physical quantities determined in such extrapolations correspond to those of an ideal solution.

5.72 A very long pipe is capped at one end with a semipermeable membrane. How deep (in meters) must the pipe be immersed into the sea for fresh water to begin passing through the membrane? Assume seawater is at 20° C and treat it as a 0.70 M NaCl solution. The density of seawater is 1.03 g cm^{-3}.

The desired process is for (fresh) water to move from a more concentrated solution (seawater) to pure solvent. This is an example of reverse osmosis, and external pressure must be provided to overcome the osmotic pressure of the seawater. The source of the pressure here is the water pressure, which increases with increasing depth. The osmotic pressure of the seawater is

$$\pi = cRT$$

$$= (0.70 \ M) \left(0.08206 \text{ L atm K}^{-1} \text{ mol}^{-1}\right) (293 \text{ K})$$

$$= 16.8 \text{ atm}$$

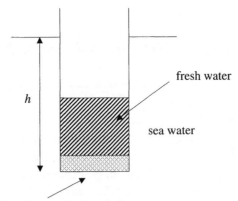

fresh water

sea water

Semipermeable membrane

The water pressure at the membrane depends on the height of the sea above it, *i.e.* the depth, $P = \rho g h$, and fresh water will begin to pass through the membrane when $P = \pi$.

$$h = \frac{\pi}{g\rho}$$

$$= \frac{16.8 \text{ atm}}{\left(9.81 \text{ m s}^{-1}\right)\left(1.03 \text{ g cm}^{-3}\right)} \left(\frac{1 \text{ m}^3}{100 \text{ cm}^3}\right)^3 \left(\frac{1000 \text{ g}}{1 \text{ kg}}\right) \left(\frac{1.01325 \times 10^5 \text{ Pa}}{1 \text{ atm}}\right)$$

$$= 168 \text{ m}$$

5.74 Calculate the solubility of $BaSO_4$ (in $g\,L^{-1}$) in **(a)** water and **(b)** a $6.5 \times 10^{-5} \, M$ $MgSO_4$ solution. The solubility product of $BaSO_4$ is 1.1×10^{-10}. Assume ideal behavior.

The equation for the dissolution of $BaSO_4$ is

$$BaSO_4(s) \rightleftharpoons Ba^{2+}(aq) + SO_4^{2-}(aq)$$

(a) If the solubility of $BaSO_4$ is $x \, M$, then there are $x \, M$ of Ba^{2+} and $x \, M \, SO_4^{2-}$ in the solution.

$$K_{sp} = 1.1 \times 10^{-10} = [Ba^{2+}][SO_4^{2-}] = x \cdot x$$

$$x = 1.05 \times 10^{-5}$$

Therefore,

$$\text{the solubility of } BaSO_4 = \left(1.05 \times 10^{-5} \text{ mol L}^{-1}\right)\left(\frac{233.4 \text{ g}}{1 \text{ mol}}\right) = 2.5 \times 10^{-3} \text{ g L}^{-1}$$

(b) The $MgSO_4$ solution contains $6.5 \times 10^{-5} \, M$ of SO_4^{2-}. If the solubility of $BaSO_4$ is $x \, M$, then there are $x \, M$ of Ba^{2+} and $x + 6.5 \times 10^{-5} \, M \, SO_4^{2-}$ in the solution.

$$K_{sp} = 1.1 \times 10^{-10} = [Ba^{2+}][SO_4^{2-}] = x \left(x + 6.5 \times 10^{-5}\right)$$

$$x^2 + 6.5 \times 10^{-5}x - 1.1 \times 10^{-10} = 0$$

$$x = 1.65 \times 10^{-6} \quad \text{or} \quad x = -6.67 \times 10^{-5} \text{ (nonphysical)}$$

Therefore,

$$\text{the solubility of BaSO}_4 = \left(1.65 \times 10^{-6} \text{ mol L}^{-1}\right)\left(\frac{233.4 \text{ g}}{1 \text{ mol}}\right) = 3.9 \times 10^{-4} \text{ g L}^{-1}$$

5.76 Oxalic acid, $(COOH)_2$, is a poisonous compound present in many plants and vegetables, including spinach. Calcium oxalate is only slightly soluble in water ($K_{sp} = 3.0 \times 10^{-9}$ at 25° C) and its ingestion can result in kidney stones. Calculate **(a)** the apparent and thermodynamic solubility of calcium oxalate in water, and **(b)** the concentrations of calcium and oxalate ions in a 0.010 M Ca(NO₃)₂ solution. Assume ideal behavior in **(b)**.

(a) The reaction corresponding to the dissolution of calcium oxalate, CaOxa, in water is

$$CaOxa(s) \rightleftharpoons Ca^{2+}(aq) + Oxa^{2-}(aq)$$

The apparent solubility is readily calculated from the given (apparent) solubility product and the stoichiometry of the dissociation. Let the solubility of CaOxa be x M, then $[Ca^{2+}] = [Oxa^{2-}] = x$.

$$K_{sp} = [Ca^{2+}][Oxa^{2-}] = x^2 = 3.0 \times 10^{-9}$$

$$x = 5.48 \times 10^{-5} M = 5.5 \times 10^{-5} M$$

The thermodynamic solubility is determined from the thermodynamic solubility product, $K_{sp}^{\circ} = \gamma_{\pm}^2 K_{sp}$, which is calculated from the apparent concentrations and the mean ionic activity obtained from the Debye–Hückel limiting law.

Since this is a very dilute solution, the molarities of the ionic species are numerically equal to the molalities required in determining the ionic strength.

$$I = \frac{1}{2}\sum_i m_i z_i^2$$

$$= \frac{1}{2}\left[\left(5.48 \times 10^{-5} m\right)(2)^2 + \left(5.48 \times 10^{-5} m\right)(2)^2\right]$$

$$= 2.19 \times 10^{-4} m$$

This is then used to determine the mean ionic activity

$$\log \gamma_{\pm} = -0.509 \left|z_+ z_-\right| \sqrt{I}$$

$$= -0.509 \left|(2)(-2)\right| \sqrt{\left(2.19 \times 10^{-4}\right)}$$

$$= -3.01 \times 10^{-2}$$

$$\gamma_{\pm} = 0.933$$

Thus, the thermodynamic solubility constant and thermodynamic solubility, x°, are

$$K_{sp}^{\circ} = (0.933)^2 \left(3.0 \times 10^{-9}\right) = 2.61 \times 10^{-9} = 2.6 \times 10^{-9} = \left(x^{\circ}\right)^2$$

$$x^{\circ} = \sqrt{2.61 \times 10^{-9}} = 5.1 \times 10^{-5} M$$

(b) The solubility in the $Ca(NO_3)_2$ solution is decreased due to the presence of Ca^{2+} ions. If the concentration of dissolved calcium oxalate is taken as x, then in the solution $[Ca^{2+}] = 0.010 + x$ and $[Oxa^{2-}] = x$.

$$K_{sp} = [Ca^{2+}][Oxa^{2-}] = 3.0 \times 10^{-9}$$

$$(0.010 + x)\,x = 3.0 \times 10^{-9}$$

$$x = 3.0 \times 10^{-7}\,M = [Oxa^{2-}]$$

$$[Ca^{2+}] = 0.010\,M + 3.0 \times 10^{-7}\,M = 0.010\,M$$

5.78 A $0.010\,m$ aqueous solution of the ionic compound $Co(NH_3)_5Cl_3$ has a freezing-point depression of 0.0558 K. What can you conclude about its structure? Assume the compound is a strong electrolyte.

The van't Hoff factor is

$$i = \frac{\Delta T}{K_f m_2} = \frac{0.0558\ \text{K}}{(1.86\ \text{K mol}^{-1}\,\text{kg})\,(0.010\ \text{mol kg}^{-1})} = 3.0$$

There are 3 particles in the solution per 1 particle before dissociation. Thus, the compound is probably $[Co(NH_3)_5Cl]Cl_2$, which on dissolution, dissociates into $[Co(NH_3)_5Cl]^{2+}$ and $2Cl^-$.

5.80 Referring to Figure 5.22, calculate the osmotic pressure for the following cases at 298 K: **(a)** The left compartment contains 200 g of hemoglobin in 1 liter of solution; the right compartment contains pure water. **(b)** The left compartment contains the same hemoglobin solution as in part **(a)**; the right compartment initially contains 6.0 g of NaCl in 1 liter of solution. Assume that the pH of the solution is such that the hemoglobin molecules are in the Na^+Hb^- form. (The molar mass of hemoglobin is 65,000 g mol^{-1}).

(a) To maintain electrical neutrality, all the Na^+ ions remain in the left compartment with the Hb^- anionic form of the protein. The total concentration is twice the hemoglobin concentration.

$$M = 2 \left[\frac{200\ \text{g}\left(\frac{1\ \text{mol}}{65000\ \text{g mol}^{-1}} \right)}{1\ \text{L}} \right] = 2\left(3.077 \times 10^{-3}\,M \right) = 6.154 \times 10^{-3}\,M$$

From the osmotic pressure equation,

$$\pi = MRT$$

$$= \left(6.154 \times 10^{-3}\,M \right)\left(0.08206\ \text{L atm K}^{-1}\,\text{mol}^{-1} \right)(298\ \text{K})$$

$$= 0.150\ \text{atm}$$

(b) Na^+ and Cl^- will diffuse through the membrane from right to left, maintaining electrical neutrality, until the chemical potentials of NaCl on both sides of the membrane are equal. According to Equation 5.71, the concentration of the NaCl, x, that diffuses from right to left depends on the concentration of NaCl initially in the right compartment, b, and the concentration of the nondiffusible ion, c.

$$b = \frac{\frac{6.0\,g}{58.44\,g\,mol^{-1}}}{1\,L} = 0.103\,M$$

$$c = 3.077 \times 10^{-3}\,M \text{ (from above)}$$

$$x = \frac{b^2}{c + 2b} = \frac{(0.103\,M)^2}{3.077 \times 10^{-3}\,M + 2\,(0.103\,M)} = 0.0507\,M$$

The osmotic pressure is determined by the difference between the number of particles in the left compartment and that in the right compartment.

$$M = \left([Hb^-] + [Na^+] + [Cl^-]\right)_L - \left([Na^+] + [Cl^-]\right)_R$$

$$= \big\{(3.077 \times 10^{-3}\,M + 3.077 \times 10^{-3}\,M + 0.0507\,M + 0.0507\,M)$$

$$- [(0.103\,M - 0.0507\,M) + (0.103\,M - 0.0507\,M)]\big\}$$

$$= 2.95 \times 10^{-3}\,M$$

Using the osmotic pressure equation,

$$\pi = MRT$$

$$= \left(2.95 \times 10^{-3}\,M\right)\left(0.08206\,L\,atm\,K^{-1}\,mol^{-1}\right)(298\,K)$$

$$= 0.072\,atm$$

5.82 The concentration of glucose inside a cell is 0.12 mM and that outside a cell is 12.3 mM. Calculate the Gibbs energy change for the transport of 3 moles of glucose into the cell at 37° C.

The process may be represented as

$$glucose(outside) \rightarrow glucose(inside)$$

This is equivalent to a concentration cell, and $\Delta G° = 0$, so

$$\Delta G = nRT \ln \frac{[glucose]_{in}}{[glucose]_{out}}$$

$$= (3\,mol)\left(8.314\,J\,K^{-1}\,mol^{-1}\right)(310.15\,K)\ln\frac{0.12\,mM}{12.3\,mM}$$

$$= -3.58 \times 10^4\,J = -35.8\,kJ$$

5.84 Use Equations 5.34 and 5.37 to calculate K_b and K_f for water shown in Table 5.2.

Using the referenced equations and the appropriate data for water,

$$K_b = \frac{RT_b^2 \mathcal{M}_1}{\Delta_{vap}\overline{H}}$$

$$= \frac{(8.314 \text{ J K}^{-1} \text{mol}^{-1}) (373.15 \text{ K})^2 (0.01802 \text{ kg mol}^{-1})}{40.79 \times 10^3 \text{ J mol}^{-1}}$$

$$= 0.5114 \text{ K mol}^{-1} \text{kg}$$

and

$$K_f = \frac{RT_f^2 \mathcal{M}_1}{\Delta_{fus}\overline{H}}$$

$$= \frac{(8.314 \text{ J K}^{-1} \text{mol}^{-1}) (273.15 \text{ K})^2 (0.01802 \text{ kg mol}^{-1})}{6.01 \times 10^3 \text{ J mol}^{-1}}$$

$$= 1.86 \text{ K mol}^{-1} \text{kg}$$

CHAPTER 6

Chemical Equilibrium

PROBLEMS AND SOLUTIONS

6.2 At 1024° C, the pressure of oxygen gas from the decomposition of copper (II) oxide (CuO) is 0.49 bar:

$$4CuO(s) \rightleftharpoons 2Cu_2O(s) + O_2(g)$$

(a) What is the value of K_P for the reaction? **(b)** Calculate the fraction of CuO that will decompose if 0.16 mole of it is placed in a 2.0-L flask at 1024° C. **(c)** What would the fraction be if a 1.0-mole sample of CuO were used? **(d)** What is the smallest amount of CuO (in moles) that would establish the equilibrium?

(a)
$$K_P = \frac{P_{O_2}}{P^o} = \frac{0.49 \text{ bar}}{1 \text{ bar}} = 0.49$$

(b) First calculate the number of moles of O_2 formed by the reaction, from which the number of moles of CuO decomposed is determined. Assume O_2 behaves ideally.

$$\text{Number of moles of } O_2 \text{ formed} = \frac{PV}{RT} = \frac{(0.49 \text{ bar}) \left(\frac{1 \text{ atm}}{1.013 \text{ bar}}\right) (2.0 \text{ L})}{(0.08206 \text{ L atm K}^{-1} \text{ mol}^{-1}) (1297 \text{ K})}$$

$$= 9.09 \times 10^{-3} \text{ mol}$$

$$\text{Number of moles of CuO decomposed} = \left(9.09 \times 10^{-3} \text{ mol } O_2\right) \left(\frac{4 \text{ mol CuO}}{1 \text{ mol } O_2}\right) = 0.0364 \text{ mol}$$

Therefore,

$$\text{Fraction of CuO decomposed} = \frac{0.0364 \text{ mol}}{0.16 \text{ mol}} = 0.23$$

(c) If a 1.0-mole sample of CuO were used, the pressure of O_2 would still be the same (0.49 bar) and it would be due to the same quantity of O_2. (A pure solid does not affect the equilibrium position, as long as it is in excess at equilibrium.) The number of moles of CuO lost would still be 0.0364 mol. Therefore,

$$\text{Fraction of CuO decomposed} = \frac{0.0364 \text{ mol}}{1.0 \text{ mol}} = 0.036$$

(d) If the number of moles of CuO were less than 0.036 mole, the equilibrium could not be established because the pressure of O_2 would be less than 0.49 bar. Therefore, the smallest number of moles of CuO needed to establish equilibrium must be slightly greater than 0.036 mole.

6.4   About 75% of the hydrogen produced for industrial use is produced by the *steam-reforming* process. This process is carried out in two stages called primary and secondary reforming. In the primary stage, a mixture of steam and methane at about 30 atm is heated over a nickel catalyst at 800° C to give hydrogen and carbon monoxide:

$$CH_4(g) + H_2O(g) \rightleftharpoons CO(g) + 3H_2(g) \qquad \Delta_r H^\circ = 206 \text{ kJ mol}^{-1}$$

The secondary stage is carried out at about 1000° C, in the presence of air, to convert the remaining methane to hydrogen:

$$CH_4(g) + \frac{1}{2}O_2(g) \rightleftharpoons CO(g) + 2H_2(g) \qquad \Delta_r H^\circ = 35.7 \text{ kJ mol}^{-1}$$

(a) What conditions of temperature and pressure would favor the formation of products in both the primary and secondary stages? (b) The equilibrium constant, K_c, for the primary stage is 18 at 800° C. (i) Calculate the value of K_P for the reaction. (ii) If the partial pressures of methane and steam were both 15 atm at the start, what would the pressures of all the gases be at equilibrium?

(a) Since both reactions are endothermic ($\Delta_r H^\circ$ is positive for each), according to Le Chatelier's principle the products would be favored at high temperatures. Indeed, the steam-reforming process is carried out at very high temperatures (between 800° C and 1000° C). It is interesting to note that in a plant that uses natural gas (methane) for both hydrogen generation and heating, about one-third of the gas is burned to maintain the high temperatures.

In each reaction there are more moles of products than reactants; therefore, products are favored at low pressures. In reality, the reactions are carried out at high pressures. The reason is that when the hydrogen gas produced is used captively (usually in the synthesis of ammonia), high pressure leads to higher yields of ammonia.

(b) (i) Using the result of Problem 6.1,

$$K_P = K_c (RT)^{\Delta n} (P^\circ)^{-\Delta n}$$

In this equation, if P° is in bar, R has to be in L bar K^{-1} mol^{-1}:

$$R = \left(0.08206 \text{ L atm K}^{-1} \text{ mol}^{-1}\right) \left(\frac{1.01325 \text{ bar}}{1 \text{ atm}}\right) = 0.08315 \text{ L bar K}^{-1} \text{ mol}^{-1}$$

and $\Delta n = (1+3) - (1+1) = 2$. Therefore,

$$K_P = (18) [(0.08315)(1073)]^2 (1)^{-2} = 1.43 \times 10^5 = 1.4 \times 10^5$$

(b) (ii) The pressures need to be converted to bars in order to use the K_P expression where P° = 1 bar. The partial pressures of the reactants are

$$(15 \text{ atm}) \left(\frac{1.013 \text{ bar}}{1 \text{ atm}}\right) = 15.2 \text{ bar}$$

The initial and equilibrium pressures of all species are shown in the following.

	CH_4	$+$	H_2O	$\rightleftharpoons$	CO	$+$	$3H_2$	
Initial	15.2		15.2		0		0	bar
At equilibrium	$15.2 - x$		$15.2 - x$		x		$3x$	bar

$$K_P = \frac{(P_{CO}/P^\circ)\,(P_{H_2}/P^\circ)^3}{(P_{CH_4}/P^\circ)\,(P_{H_2O}/P^\circ)}$$

$$1.43 \times 10^5 = \frac{x\,(3x)^3}{(15.2 - x)^2} = \frac{27x^4}{(15.2 - x)^2}$$

Take the square root of K_P and the last term in the above expression.

$$378 = \frac{5.20x^2}{15.2 - x}$$

$$5.20x^2 + 378x - 5745.6 = 0$$

$$x = 12.9 \quad \text{or} \quad -85.6 \text{ (nonphysical)}$$

Therefore, at equilibrium,

$$P_{CH_4} = (15.2 - x) \text{ bar} = 2 \text{ bar} = 2 \text{ atm}$$

$$P_{H_2O} = (15.2 - x) \text{ bar} = 2 \text{ bar} = 2 \text{ atm}$$

$$P_{CO} = x \text{ bar} = 13 \text{ bar} = 13 \text{ atm}$$

$$P_{H_2} = 3x \text{ bar} = 39 \text{ bar} = 38 \text{ atm}$$

6.6 The vapor pressure of mercury is 0.002 mmHg at 26° C. **(a)** Calculate the values of K_c and K_P for the process $Hg(l) \rightleftharpoons Hg(g)$. **(b)** A chemist breaks a thermometer and spills mercury onto the floor of a laboratory measuring 6.1 m long, 5.3 m wide, and 3.1 m high. Calculate the mass of mercury (in grams) vaporized at equilibrium and the concentration of mercury vapor in mg m^{-3}. Does this concentration exceed the safety limit of 0.05 mg m^{-3}? (Ignore the volume of furniture and other objects in the laboratory.)

(a)

$$K_P = \frac{P_{Hg}}{P^\circ} = \frac{(0.002 \text{ mmHg}) \left(\frac{1 \text{ atm}}{760 \text{ mmHg}}\right) \left(\frac{1.013 \text{ bar}}{1 \text{ atm}}\right)}{1 \text{ bar}} = 2.666 \times 10^{-6} = 2.67 \times 10^{-6}$$

Using the result derived in Problem 6.1,

$$K_c = \frac{K_P}{(RT)^{\Delta n}\,(P^\circ)^{-\Delta n}}$$

To use this expression, R has to be expressed in L bar K^{-1} mol^{-1}, which has been done in Problem 6.4. Δn is 1.

$$K_c = \frac{2.666 \times 10^{-6}}{[(0.08315)\,(299.2)]^1\,(1)^{-1}} = 1.07 \times 10^{-7}$$

(b) The volume occupied by Hg vapor is the same as the volume of the room.

$$V = (6.1 \text{ m}) \ (5.3 \text{ m}) \ (3.1 \text{ m}) = \left(100 \text{ m}^3\right) \left(\frac{1000 \text{ L}}{1 \text{ m}^3}\right) = 1.00 \times 10^5 \text{ L}$$

At equilibrium, the vapor pressure of Hg is 0.002 mmHg. The number of moles of Hg vapor is calculated by using the ideal gas law.

$$n = \frac{PV}{RT} = \frac{(0.002 \text{ mmHg}) \left(\frac{1 \text{ atm}}{760 \text{ mmHg}}\right) \left(1.00 \times 10^5 \text{ L}\right)}{(0.08206 \text{ L atm K}^{-1} \text{ mol}^{-1}) \ (299.2 \text{ K})} = 1.07 \times 10^{-2} \text{ mol}$$

Therefore, the mass of Hg vapor is

$$m = \left(1.07 \times 10^{-2} \text{ mol}\right) \left(200.6 \text{ g mol}^{-1}\right) = 2.15 \text{ g}$$

and the concentration of Hg is

$$\text{Concentration} = \frac{2.15 \text{ g}}{100 \text{ m}^3} = 2.2 \times 10^{-2} \text{ g m}^{-3} = 22 \text{ mg m}^{-3}$$

This concentration greatly exceeds the safety limit.

6.8   Consider the thermal decomposition of $CaCO_3$:

$$CaCO_3(s) \rightleftharpoons CaO(s) + CO_2(g)$$

The equilibrium vapor pressures of CO_2 are 22.6 mmHg at 700° C and 1829 mmHg at 950° C. Calculate the standard enthalpy of the reaction.

The van't Hoff equation,

$$\ln \frac{K_2}{K_1} = \frac{\Delta_r H^\circ}{R} \left(\frac{1}{T_1} - \frac{1}{T_2}\right),$$

is used to solve this problem. Since

$$K_P = \frac{P_{CO_2}}{P^\circ},$$

K_P is proportional to P_{CO_2}. Thus,

$$\ln \frac{K_2}{K_1} = \ln \frac{P_{CO_2,2}}{P_{CO_2,1}} = \frac{\Delta_r H^\circ}{R} \left(\frac{1}{T_1} - \frac{1}{T_2}\right)$$

$$\ln \frac{1829 \text{ mmHg}}{22.6 \text{ mmHg}} = \frac{\Delta_r H^\circ}{8.314 \text{ J K}^{-1} \text{ mol}^{-1}} \left(\frac{1}{973.2 \text{ K}} - \frac{1}{1223.2 \text{ K}}\right)$$

$$\Delta_r H^\circ = 1.74 \times 10^5 \text{ J mol}^{-1}$$

6.10   The vapor pressure of dry ice (solid CO_2) is 672.2 torr at $-80°$ C and 1486 torr at $-70°$ C. Calculate the molar heat of sublimation of CO_2.

The van't Hoff equation

$$\ln \frac{K_2}{K_1} = \frac{\Delta_r H^\circ}{R} \left(\frac{1}{T_1} - \frac{1}{T_2} \right)$$

is used to solve this problem. The process is

$$CO_2(s) \rightleftharpoons CO_2(g),$$

and the equilibrium constant is

$$K_P = \frac{P_{CO_2}}{P^\circ}.$$

Since K_P is proportional to P_{CO_2}, the van't Hoff equation becomes

$$\ln \frac{K_2}{K_1} = \ln \frac{P_{CO_2,2}}{P_{CO_2,1}} = \frac{\Delta_r H^\circ}{R} \left(\frac{1}{T_1} - \frac{1}{T_2} \right)$$

$$\ln \frac{1486 \text{ torr}}{672.2 \text{ torr}} = \frac{\Delta_r H^\circ}{8.314 \text{ J K}^{-1} \text{mol}^{-1}} \left(\frac{1}{193.2 \text{ K}} - \frac{1}{203.2 \text{ K}} \right)$$

$$\Delta_r H^\circ = 2.59 \times 10^4 \text{ J mol}^{-1}$$

6.12   Calculate the value of $\Delta_r G^\circ$ for each of the following equilibrium constants: 1.0×10^{-4}, 1.0×10^{-2}, 1.0, 1.0×10^2, 1.0×10^4 at 298 K.

$\Delta_r G^\circ$ can be evaluated using $\Delta_r G^\circ = -RT \ln K$.

K	$\Delta_r G^\circ/\text{J·mol}^{-1}$
1.0×10^{-4}	2.3×10^4
1.0×10^{-2}	1.1×10^4
1.0	0
1.0×10^2	-1.1×10^4
1.0×10^4	-2.3×10^4

6.14   The dissociation of N_2O_4 into NO_2 is 16.7% complete at 298 K and 1 atm:

$$N_2O_4(g) \rightleftharpoons 2NO_2(g)$$

Calculate the equilibrium constant and the standard Gibbs energy change for the reaction. [*Hint*: Let α be the degree of dissociation and show that $K_P = 4\alpha^2 P/(1 - \alpha^2)$, where P is the total pressure.]

The equilibrium constant is related to the degree of dissociation, α. The exact relation is derived as follows.

The dissociation reaction is

$$N_2O_4(g) \rightleftharpoons 2NO_2(g)$$

Let the initial number of moles of N_2O_4 be n. At equilibrium, $n\alpha$ moles of N_2O_4 dissociates, giving $2n\alpha$ moles of NO_2. The number of moles of N_2O_4 remaining is $n - n\alpha = n(1 - \alpha)$. The total number of moles of gases are $n(1 - \alpha) + 2n\alpha = n(1 + \alpha)$.

The equilibrium constant can be calculated once the partial pressures of the gases are known. The partial pressures are calculated from the mole fractions of the gases.

$$x_{N_2O_4} = \frac{n(1-\alpha)}{n(1+\alpha)} = \frac{1-\alpha}{1+\alpha}$$

$$x_{NO_2} = \frac{2n\alpha}{n(1+\alpha)} = \frac{2\alpha}{1+\alpha}$$

The partial pressures are

$$P_{N_2O_4} = x_{N_2O_4}P = \frac{1-\alpha}{1+\alpha}P$$

$$P_{NO_2} = x_{NO_2}P = \frac{2\alpha}{1+\alpha}P$$

The equilibrium constant can now be calculated using the partial pressures in bars. The total pressure is 1 atm, which is equivalent to 1.013 bar.

$$K_P = \frac{P_{NO_2}^2}{P_{N_2O_4}} = \frac{\left(\frac{2\alpha}{1+\alpha}P\right)^2}{\frac{1-\alpha}{1+\alpha}P} = \frac{4\alpha^2 P}{(1+\alpha)(1-\alpha)} = \frac{4\alpha^2}{1-\alpha^2}P = \frac{4(0.167)^2}{1-0.167^2}(1.013)$$

$$= 0.1162 = 0.116$$

The standard Gibbs energy change is

$$\Delta_r G^\circ = -RT \ln K = -\left(8.314\,\text{J}\,\text{K}^{-1}\,\text{mol}^{-1}\right)(298\,\text{K})\ln 0.1162 = 5.33 \times 10^3\,\text{J}\,\text{mol}^{-1}$$

6.16   Consider the decomposition of magnesium carbonate:

$$MgCO_3(s) \rightleftharpoons MgO(s) + CO_2(g)$$

Calculate the temperature at which the decomposition begins to favor products. Assume that $\Delta_r H^\circ$ and $\Delta_r S^\circ$ are temperature independent. Use the data in Appendix 2 for your calculation.

As discussed in the text for the decomposition of calcium carbonate, a reaction favors the formation of products at equilbrium when

$$\Delta_r G^\circ = \Delta_r H^\circ - T\Delta_r S^\circ < 0$$

$\Delta_r H^\circ$ and $\Delta_r S^\circ$ are

$$\Delta_r H^\circ = \Delta_f \overline{H}^\circ \left[MgO(s) \right] + \Delta_f \overline{H}^\circ \left[CO_2(g) \right] - \Delta_f \overline{H}^\circ \left[MgCO_3(s) \right]$$

$$= -601.8 \text{ kJ mol}^{-1} + \left(-393.5 \text{ kJ mol}^{-1} \right) - \left(-1095.8 \text{ kJ mol}^{-1} \right)$$

$$= 100.5 \text{ kJ mol}^{-1}$$

$$\Delta_r S^\circ = \overline{S}^\circ \left[MgO(s) \right] + \overline{S}^\circ \left[CO_2(g) \right] - \overline{S}^\circ \left[MgCO_3(s) \right]$$

$$= 26.78 \text{ J K}^{-1} \text{mol}^{-1} + 213.6 \text{ J K}^{-1} \text{mol}^{-1} - 65.7 \text{ J K}^{-1} \text{mol}^{-1}$$

$$= 174.68 \text{ J K}^{-1} \text{mol}^{-1}$$

Therefore, for the reaction to begin to favor products,

$$\Delta_r H^\circ - T \Delta_r S^\circ = 100.5 \times 10^3 \text{ J mol}^{-1} - T \left(174.68 \text{ J K}^{-1} \text{mol}^{-1} \right) < 0$$

$$T > \frac{100.5 \times 10^3 \text{ J mol}^{-1}}{174.68 \text{ J K}^{-1} \text{mol}^{-1}}$$

$$T > 575.3 \text{ K}$$

6.18 Consider the reaction

$$2NO_2(g) \rightleftharpoons N_2O_4(g) \qquad \Delta_r H^\circ = -58.04 \text{ kJ mol}^{-1}$$

Predict what happens to the system at equilibrium if **(a)** the temperature is raised, **(b)** the pressure on the system is increased, **(c)** an inert gas is added to the system at constant pressure, **(d)** an inert gas is added to the system at constant volume, and **(e)** a catalyst is added to the system.

(a) The equilibrium will shift from right to left. The equilibrium constant K_P will decrease.

(b) The equilibrium will shift from left to right. K_P remains unchanged.

(c) The volume of the system expands, and the gases are "diluted." K_P remains unchanged, but the equilibrium shifts from right to left.

(d) The pressure of the system will increase, but the partial pressures of NO_2 and N_2O_4 remain constant. K_P remains unchanged, and the position of equilibrium will not shift.

(e) The catalyst has no effect on either K_P or the position of equilibrium.

6.20 At a certain temperature, the equilibrium pressures of NO_2 and N_2O_4 are 1.6 bar and 0.58 bar, respectively. If the volume of the container is doubled at constant temperature, what would be the partial pressures of the gases when equilibrium is re-established?

The equilibrium process is

$$N_2O_4(g) \rightleftharpoons 2NO_2(g)$$

The equilibrium constant can be calculated using the equilibrium pressures.

$$K_P = \frac{1.6^2}{0.58} = 4.41$$

When the volume of the container is doubled at constant temperature, the partial pressures of the gases are halved. That is,

$$P_{NO_2} = 0.80 \text{ bar}$$

$$P_{N_2O_4} = 0.29 \text{ bar}$$

According to Le Chatelier's principle, when the volume is increased, the reaction will shift to produce more molecules. In the case at hand, more NO_2 will be produced.

Assume the equilibrium pressure of N_2O_4 to be $0.29 - x$ bars. The equilibrium pressure of NO_2 is obtained using the stoichiometric relationship between the two compounds.

	N_2O_4	$\rightleftharpoons$	$2NO_2$	
At equilibrium	$0.29 - x$		$0.80 + 2x$	bar

Solve for x using the equilibrium expression and discarding the non-physical root.

$$K_P = 4.41 = \frac{(0.80 + 2x)^2}{0.29 - x}$$

$$4.41\,(0.29 - x) = (0.80 + 2x)^2$$

$$4x^2 + 7.61x - 0.639 = 0$$

$$x = 8.06 \times 10^{-2}$$

Therefore, when equilibrium is re-established,

$$P_{N_2O_4} = 0.29 - x = 0.21 \text{ bar}$$

$$P_{NO_2} = 0.80 + 2x = 0.96 \text{ bar}$$

6.22 Photosynthesis can be represented by

$$6CO_2(g) + 6H_2O(l) \rightleftharpoons C_6H_{12}O_6(s) + 6O_2(g) \qquad \Delta_r H^\circ = 2801 \text{ kJ mol}^{-1}$$

Explain how the equilibrium would be affected by the following changes: **(a)** the partial pressure of CO_2 is increased, **(b)** O_2 is removed from the mixture, **(c)** $C_6H_{12}O_6$ (glucose) is removed from the mixture, **(d)** more water is added, **(e)** a catalyst is added, **(f)** the temperature is decreased, and **(g)** more sunlight shines on the plants.

(a) The equilibrium would shift from left to right.

(b) The equilibrium would shift from left to right.

(c) The equilibrium would be unaffected, since $C_6H_{12}O_6$ is a solid. (As long as excess solid remains.)

(d) The equilibrium would be unaffected, since water is a liquid. (As long as excess water remains.)

(e) The catalyst has no effect on the position of equilibrium.

(f) The equilibrium would shift from right to left.

(g) Assuming constant temperature, the position of equilibrium is unaffected by the amount of sunlight shining on the plant, although the rate of attaining equilibrium is increased.

6.24 Industrially, sodium metal is obtained by electrolyzing molten sodium chloride. The reaction at the cathode is $Na^+ + e^- \rightarrow Na$. We might expect that potassium metal could also be prepared by electrolyzing molten potassium chloride. Potassium metal is soluble in molten potassium chloride, however, and is therefore hard to recover. Furthermore, potassium vaporizes readily at the operating temperature, creating hazardous conditions. Instead, potassium is prepared by the distillation of molten potassium chloride in the presence of sodium vapor at 892° C:

$$Na(g) + KCl(l) \rightleftharpoons NaCl(l) + K(g)$$

Considering that potassium is a stronger reducing agent than sodium, explain why this approach works. (The boiling points of sodium and potassium are 892° C and 770° C, respectively.)

Potassium is more volatile than sodium and is removed from the system more rapidly, causing the equilibrium to shift from left to right.

6.26 Derive Equation 6.23 from 6.21.

$$Y = \frac{[L]}{[L] + K_d}$$

$$Y\left([L] + K_d\right) = [L]$$

$$Y[L] + YK_d = [L]$$

$$\frac{Y[L] + YK_d}{[L]} = 1$$

$$Y + \frac{Y}{[L]}K_d = 1$$

$$\frac{Y}{[L]}K_d = 1 - Y$$

$$\frac{Y}{[L]} = \frac{1}{K_d} - \frac{Y}{K_d}$$

6.28 An equilibrium dialysis experiment showed that the concentrations of the free ligand, bound ligand, and protein are $1.2 \times 10^{-5}\,M$, $5.4 \times 10^{-6}\,M$, and $4.9 \times 10^{-6}\,M$, respectively. Calculate the dissociation constant for the reaction $PL \rightleftharpoons P + L$. Assume there is one binding site per protein molecule.

Since there is one binding site per protein molecule, $[PL] = [L]_{bound}$.

$$K_d = \frac{[P][L]}{[PL]} = \frac{\left(4.9 \times 10^{-6}\right)\left(1.2 \times 10^{-5}\right)}{5.4 \times 10^{-6}} = 1.1 \times 10^{-5}$$

6.30   As mentioned in the chapter, the standard Gibbs energy for the hydrolysis of ATP to ADP at 310 K is approximately $-30.5 \text{ kJ mol}^{-1}$. Calculate the value of $\Delta_r G^{o\prime}$ for the reaction in the muscle of a polar sea fish at $-1.5°$ C. (*Hint*: $\Delta_r H^{o\prime} = -20.1 \text{ kJ mol}^{-1}$.)

$\Delta_r S^{o\prime}$ can be determined from $\Delta_r G^{o\prime}$ at 310 K and $\Delta_r H^{o\prime}$, which is assumed to be temperature independent.

$$\Delta_r G^{o\prime} = \Delta_r H^{o\prime} - T \Delta_r S^{o\prime}$$

$$\Delta_r S^{o\prime} = \frac{\Delta_r H^{o\prime} - \Delta_r G^{o\prime}}{T} = \frac{-20.1 \text{ kJ mol}^{-1} - \left(-30.5 \text{ kJ mol}^{-1}\right)}{310 \text{ K}} = 3.355 \times 10^{-2} \text{ kJ K}^{-1} \text{mol}^{-1}$$

Assuming that $\Delta_r S^{o\prime}$ is temperature independent, in addition to $\Delta_r H^{o\prime}$, $\Delta_r G^{o\prime}$ at $-1.5°$ C (271.65 K) can be determined.

$$\Delta_r G^{o\prime} = \Delta_r H^{o\prime} - T \Delta_r S^{o\prime} = -20.1 \text{ kJ mol}^{-1} - (271.65 \text{ K}) \left(3.355 \times 10^{-2} \text{ kJ K}^{-1} \text{mol}^{-1}\right)$$

$$= -29.2 \text{ kJ mol}^{-1}$$

6.32   The formation of a dipeptide is the first step toward the synthesis of a protein molecule. Consider the following reaction:

$$\text{glycine} + \text{glycine} \rightarrow \text{glycylglycine} + H_2O$$

Use the data in Appendix 2 to calculate the value of $\Delta_r G^{o\prime}$ and the equilibrium constant at 298 K, keeping in mind that the reaction is carried out in an aqueous buffer solution. Assume that the value of $\Delta_r G^{o\prime}$ is essentially the same at 310 K. What conclusion can you draw about your result?

$\Delta_r G^{o\prime}$ is calculated from the standard molar Gibbs energies of formation of the reactants and products.

$$\Delta_r G^{o\prime} = \Delta_f \overline{G}^o \left(\text{glycylglycine}\right) + \Delta_f \overline{G}^o \left(H_2O\right) - 2\Delta_f \overline{G}^o \left(\text{glycine}\right)$$

$$= \left(-493.1 \text{ kJ mol}^{-1}\right) + \left(-237.2 \text{ kJ mol}^{-1}\right) - 2 \left(-379.9 \text{ kJ mol}^{-1}\right)$$

$$= 29.5 \text{ kJ mol}^{-1}$$

The equilibrium constant can now be determined.

$$\ln K' = -\frac{\Delta_r G^{o\prime}}{RT} = -\frac{29.5 \times 10^3 \text{ J mol}^{-1}}{\left(8.314 \text{ J K}^{-1} \text{mol}^{-1}\right)(298 \text{ K})} = -11.91$$

$$K' = 6.72 \times 10^{-6}$$

If $\Delta_r G^{o\prime}$ at 310 K is essentially the same as that at 298 K, then K' at 310 K is also approximately the same as that at 298 K. The small K' indicates that the formation of a dipeptide (and hence a protein molecule) is not a spontaneous process under standard-state conditions. Protein synthesis *in vivo* is carried out both under other conditions and with the aid of ATP.

6.34 A polypeptide can exist in either the helical or random coil forms. The equilibrium constant for the equilibrium reaction of the helix to the random coil transition is 0.86 at 40° C and 0.35 at 60° C. Calculate the values of $\Delta_r H°$ and $\Delta_r S°$ for the reaction.

$\Delta_r H°$ is calculated from the van't Hoff equation.

$$\ln \frac{K_2}{K_1} = \frac{\Delta_r H°}{R} \left(\frac{1}{T_1} - \frac{1}{T_2} \right)$$

$$\ln \frac{0.35}{0.86} = \frac{\Delta_r H°}{8.314 \, \text{J K}^{-1} \, \text{mol}^{-1}} \left(\frac{1}{313 \, \text{K}} - \frac{1}{333 \, \text{K}} \right)$$

$$\Delta_r H° = -3.90 \times 10^4 \, \text{J mol}^{-1} = -3.9 \times 10^4 \, \text{J mol}^{-1}$$

To calculate $\Delta_r S°$, $\Delta_r G°$ at a particular temperature is needed. The following calculations are carried out using 40° C.

$$\Delta_r G° = -RT \ln K = - \left(8.314 \, \text{J K}^{-1} \, \text{mol}^{-1} \right) (313 \, \text{K}) \ln 0.86 = 392 \, \text{J mol}^{-1}$$

Assuming $\Delta_r H°$ and $\Delta_r S°$ to be independent of temperature, the latter can be determined.

$$\Delta_r G° = \Delta_r H° - T \Delta_r S°$$

$$\Delta_r S° = \frac{\Delta_r H° - \Delta_r G°}{T} = \frac{-3.90 \times 10^4 \, \text{J mol}^{-1} - 392 \, \text{J mol}^{-1}}{313 \, \text{K}} = -1.3 \times 10^2 \, \text{J K}^{-1} \, \text{mol}^{-1}$$

6.36 At a certain temperature, the equilibrium partial pressures are $P_{\text{NH}_3} = 321.6$ atm, $P_{\text{N}_2} = 69.6$ atm, and $P_{\text{H}_2} = 208.8$ atm, respectively. **(a)** Calculate the value of K_P for the reaction described in Example 6.1. **(b)** Calculate the thermodynamic equilibrium constant if $\gamma_{\text{NH}_3} = 0.782$, $\gamma_{\text{N}_2} = 1.266$, and $\gamma_{\text{H}_2} = 1.243$.

The reaction is

$$N_2(g) + 3H_2(g) \rightleftharpoons 2NH_3(g)$$

(a) The partial pressures have to be expressed in bars to be used in the equilibrium expression.

$$P_{\text{NH}_3} = (321.6 \, \text{atm}) \left(\frac{1.01325 \, \text{bar}}{1 \, \text{atm}} \right) = 325.86 \, \text{bar}$$

$$P_{\text{N}_2} = (69.6 \, \text{atm}) \left(\frac{1.01325 \, \text{bar}}{1 \, \text{atm}} \right) = 70.52 \, \text{bar}$$

$$P_{\text{H}_2} = (208.8 \, \text{atm}) \left(\frac{1.01325 \, \text{bar}}{1 \, \text{atm}} \right) = 211.57 \, \text{bar}$$

The equilibrium constant is

$$K_P = \frac{P_{\text{NH}_3}^2}{P_{\text{N}_2} P_{\text{H}_2}^3} = \frac{(325.86)^2}{(70.52)(211.57)^3} = 1.590 \times 10^{-4} = 1.59 \times 10^{-4}$$

(b) $K_f = K_\gamma K_P = \dfrac{\gamma_{NH_3}^2}{\gamma_{N_2}\gamma_{H_2}^3} K_P = \dfrac{(0.782)^2}{(1.266)(1.243)^3}\left(1.590 \times 10^{-4}\right) = 4.00 \times 10^{-5}$

6.38   The solubility of n-heptane in water is 0.050 g per liter of solution at 25° C. What is the Gibbs energy change for the hypothetical process of dissolving n-heptane in water at a concentration of 2.0 g L^{-1} at the same temperature? (*Hint*: First calculate the value of $\Delta_r G^\circ$ from the equilibrium process and then the $\Delta_r G$ value using Equation 6.6.)

The solubility of n-heptane in water is $\dfrac{0.050 \text{ g L}^{-1}}{100.20 \text{ g mol}^{-1}} = 4.99 \times 10^{-4}$ M. The appropriate equilibrium is

$$n\text{-heptane}(l) \rightleftharpoons n\text{-heptane}(aq)$$

with $K_{sp}[n\text{-heptane}] = 4.99 \times 10^{-4}$. For the saturated solution, $\Delta_r G = 0$, and

$$\Delta_r G^\circ = -RT \ln K$$

$$= -\left(8.314 \text{ J K}^{-1} \text{mol}^{-1}\right)(298 \text{ K}) \ln\left(4.99 \times 10^{-4}\right)$$

$$= 1.88 \times 10^4 \text{ J mol}^{-1}$$

For the hypothetical process, the concentration is $c = \dfrac{2.0 \text{ g L}^{-1}}{100.20 \text{ g mol}^{-1}} = 2.00 \times 10^{-2}$ M, and

$$\Delta_r G = \Delta_r G^\circ + RT \ln c$$

$$= 1.88 \times 10^4 \text{ J mol}^{-1} + \left(8.314 \text{ J K}^{-1}\text{mol}^{-1}\right)(298 \text{ K}) \ln\left(2.00 \times 10^{-2}\right)$$

$$= 9.14 \times 10^3 \text{ J mol}^{-1}$$

$$= 9.1 \text{ kJ mol}^{-1}$$

6.40   The binding of oxygen to hemoglobin (Hb) is quite complex, but for our purpose we can represent the reaction as

$$Hb(aq) + O_2(g) \rightarrow HbO_2(aq)$$

If the value of $\Delta_r G^\circ$ for the reaction is -11.2 kJ mol^{-1} at 20° C, calculate the value of $\Delta_r G^{\circ\prime}$ for the reaction. (*Hint*: Refer to the result in Problem 6.39.)

Using 1 bar and 1 M as standard states for gases and solutions,

$$\Delta_r G = \Delta_r G^\circ + RT \ln \frac{[HbO_2]/1\, M}{([Hb]/1\, M)\left(P_{O_2}/1 \text{ bar}\right)}$$

Using the biochemical standard states,

$$\Delta_r G = \Delta_r G^{\circ\prime} + RT \ln \frac{[HbO_2]/1\, M}{([Hb]/1\, M)\left(P_{O_2}/0.2 \text{ bar}\right)}$$

Since the value of $\Delta_r G$ does not depend on the standard states chosen, the two expressions above can be equated.

$$\Delta_r G^{\circ\prime} + RT \ln \frac{[HbO_2]/1\,M}{([Hb]/1\,M)\,(P_{O_2}/0.2\,\text{bar})} = \Delta_r G^{\circ} + RT \ln \frac{[HbO_2]/1\,M}{([Hb]/1\,M)\,(P_{O_2}/1\,\text{bar})}$$

$$\Delta_r G^{\circ\prime} = \Delta_r G^{\circ} + RT \ln \frac{1\,\text{bar}}{0.2\,\text{bar}}$$

$$\Delta_r G^{\circ\prime} = -11.2 \times 10^3\,\text{J mol}^{-1} + \left(8.314\,\text{J K}^{-1}\,\text{mol}^{-1}\right)(293.2\,\text{K}) \ln \frac{1}{0.2} = -7.3 \times 10^3\,\text{J mol}^{-1}$$

6.42 Many hydrocarbons exist as structural isomers, which are compounds that have the same molecular formula but different structures. For example, both butane and isobutane have the same molecular formula: C_4H_{10}. Calculate the mole percent of these molecules in an equilibrium mixture at 25° C, given that the standard Gibbs energy of formation of butane is $-15.9\,\text{kJ mol}^{-1}$ and that of isobutane is $-18.0\,\text{kJ mol}^{-1}$. Does your result support the notion that straight-chain hydrocarbons (that is, hydrocarbons in which the C atoms are joined in a line) are less stable than branch-chain hydrocarbons?

The isomerization process can be expressed as

$$\text{butane}(g) \rightleftharpoons \text{isobutane}(g)$$

First calculate $\Delta_r G^{\circ}$, from which the equilibrium constant K_P is obtained.

$$\Delta_r G^{\circ} = \Delta_f \overline{G}^{\circ}\,[\text{isobutane}] - \Delta_f \overline{G}^{\circ}\,[\text{butane}]$$

$$= -18.0\,\text{kJ mol}^{-1} - \left(-15.9\,\text{kJ mol}^{-1}\right)$$

$$= -2.1\,\text{kJ mol}^{-1}$$

$$\ln K_P = -\frac{\Delta_r G^{\circ}}{RT}$$

$$= -\frac{-2.1 \times 10^3\,\text{J mol}^{-1}}{\left(8.314\,\text{J K}^{-1}\,\text{mol}^{-1}\right)(298\,\text{K})} = 0.848$$

$$K_P = 2.33$$

The equilibrium constant is a ratio between the pressures of isobutane and butane. The pressure of each gas is proportional to number of moles of gas at constant T and V. Therefore,

$$K_P = \frac{P_{\text{isobutane}}}{P_{\text{butane}}} = \frac{n_{\text{isobutane}}}{n_{\text{butane}}} = 2.33$$

According to the above expression, for each mole of butane, there are 2.33 moles of isobutane.

$$\text{Mole percent of isobutane} = \frac{2.33}{3.33} \times 100\% = 70\%$$

$$\text{Mole percent of butane} = 1 - 70\% = 30\%$$

These results support the notion that straight-chain hydrocarbons like butane are less stable than branch-chain hydrocarbons like isobutane.

6.44   Comment on the validity of using concentrations instead of activities when discussing reactions in biological cells.

Since concentrations in biological cells are typically low, it is valid to use them instead of activities.

6.46   The following data show the binding of Mg^{2+} ions with a protein containing n equivalent sites:

$[Mg^{2+}]_{total}/\mu M$	108	180	288	501	752
$[Mg^{2+}]_{free}/\mu M$	35	65	115	248	446

Apply the Scatchard plot to determine n and K_d. The protein concentration is 98 μM.

The Scatchard plot follows from the equation

$$\frac{Y}{[L]} = \frac{n}{K_d} - \frac{Y}{K_d}$$

and is a plot of $\frac{Y}{[L]}$ vs Y, where

$$Y = \frac{\left[Mg^{2+}\right]_{bound}}{[P]}$$

$$[L] = \left[Mg^{2+}\right]_{free} = \left[Mg^{2+}\right]_{total} - \left[Mg^{2+}\right]_{bound}$$

The slope of the plot is $-\frac{1}{K_d}$ and the intercept is $\frac{n}{K_d}$. The relevant data are given in the table below.

$[Mg^{2+}]_{total}/\mu M$	108	180	288	501	752
$[Mg^{2+}]_{free}/\mu M$	35	65	115	248	446
$[Mg^{2+}]_{bound}/\mu M$	73	115	173	253	306
Y	0.745	1.173	1.765	2.582	3.122
$\frac{Y}{[Mg^{2+}]_{free}}/10^4 \cdot M^{-1}$	2.13	1.80	1.53	1.04	0.700

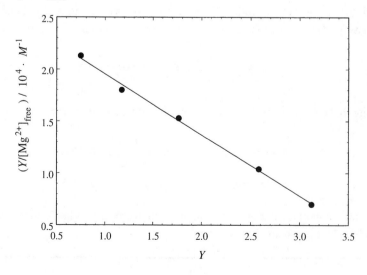

The equation for the best fit line to the data is $y = -5.86 \times 10^3 x + 2.54 \times 10^4$. Therefore,

$$K_d = -\frac{1}{\text{slope}} = -\frac{1}{-5.86 \times 10^3} = 1.71 \times 10^{-4} = 1.7 \times 10^{-4}$$

$$n = (\text{intercept})\,(K_d) = \left(2.54 \times 10^4\right)\left(1.71 \times 10^{-4}\right) = 4.34 \approx 4$$

CHAPTER 7

Electrochemistry

PROBLEMS AND SOLUTIONS

7.2 Calculate the emf of the Daniell cell at 298 K when the concentrations of $CuSO_4$ and $ZnSO_4$ are 0.50 *M* and 0.10 *M*, respectively. What would the emf be if activities were used instead of concentrations? (The $\gamma_\pm$ values for $CuSO_4$ and $ZnSO_4$ at their respective concentrations are 0.068 and 0.15, respectively.)

The half reactions for the Daniell cell are

$$\text{Anode:} \quad \mathrm{Zn} \rightarrow \mathrm{Zn^{2+}} + 2\mathrm{e^-}$$

$$\text{Cathode:} \quad \mathrm{Cu^{2+}} + 2\mathrm{e^-} \rightarrow \mathrm{Cu}$$

Thus, for the cell,

$$E^\circ = 0.342\ \text{V} - (-0.762\ \text{V}) = 1.104\ \text{V}$$

The emf at the specified Cu^{2+} and Zn^{2+} concentrations is

$$E = E^\circ - \frac{0.0257\ \text{V}}{\nu} \ln \frac{[\mathrm{Zn^{2+}}]}{[\mathrm{Cu^{2+}}]} = 1.104\ \text{V} - \frac{0.0257\ \text{V}}{2} \ln \frac{0.10}{0.50} = 1.125\ \text{V}$$

Using activities,

$$\begin{aligned} E &= E^\circ - \frac{0.0257\ \text{V}}{\nu} \ln \frac{a_{\mathrm{Zn^{2+}}}}{a_{\mathrm{Cu^{2+}}}} \\ &= E^\circ - \frac{0.0257\ \text{V}}{\nu} \ln \frac{\gamma_{\pm,\mathrm{ZnSO_4}}[\mathrm{Zn^{2+}}]}{\gamma_{\pm,\mathrm{CuSO_4}}[\mathrm{Cu^{2+}}]} \\ &= 1.104\ \text{V} - \frac{0.0257\ \text{V}}{2} \ln \frac{(0.15)(0.10)}{(0.068)(0.50)} \\ &= 1.115\ \text{V} \end{aligned}$$

7.4 Consider a Daniell cell operating under non-standard-state conditions. Suppose that the cell's reaction is multiplied by 2. What effect does this have on each of the following quantities in the Nernst equation? **(a)** E, **(b)** E°, **(c)** Q, **(d)** $\ln Q$, and **(e)** ν

(a) None. E is an intensive property, (b) none. E° is an intensive property, (c) squared, (d) doubled, (e) doubled.

7.6 From the standard reduction potentials listed in Table 7.1 for $Cu^{2+} \mid Cu$ and $Pt \mid Cu^{2+}, Cu^+$, calculate the standard reduction potential for $Cu^+ \mid Cu$.

The reduction half reactions and the standard reduction potentials for $Cu^{2+} \mid Cu$ and $Pt \mid Cu^{2+}, Cu^+$ are

$$Cu^{2+} + 2e^- \rightarrow Cu \qquad E_1^o = 0.342 \text{ V}$$

$$Cu^{2+} + e^- \rightarrow Cu^+ \qquad E_2^o = 0.153 \text{ V}$$

The reduction half reaction for $Cu^+ \mid Cu$ is

$$Cu^+ + e^- \rightarrow Cu.$$

Let its reduction potential be denoted as E_3^o. This half reaction can be derived from the previous two half reactions. The Gibbs energy changes are related by

$$\Delta_r G_3^o = \Delta_r G_1^o - \Delta_r G_2^o$$

Since $\Delta_r G^\circ = -\nu F E^\circ$, the above relation becomes

$$\left(-\nu_3 F E_3^o \right) = \left(-\nu_1 F E_1^o \right) - \left(-\nu_2 F E_2^o \right)$$

$$\nu_3 E_3^o = \nu_1 E_1^o - \nu_2 E_2^o$$

$$E_3^o = \frac{\nu_1 E_1^o - \nu_2 E_2^o}{\nu_3}$$

$$= \frac{2 \, (0.342 \text{ V}) - 1 \, (0.153 \text{ V})}{1} = 0.531 \text{ V}$$

7.8 Calculate the values of E°, $\Delta_r G^\circ$, and K for the following reactions at 25° C:

(a) $Zn + Sn^{4+} \rightleftharpoons Zn^{2+} + Sn^{2+}$

(b) $Cl_2 + 2I^- \rightleftharpoons 2Cl^- + I_2$

(c) $5Fe^{2+} + MnO_4^- + 8H^+ \rightleftharpoons Mn^{2+} + 4H_2O + 5Fe^{3+}$

(a)   Anode: $Zn \rightarrow Zn^{2+} + 2e^-$

                             Cathode: $Sn^{4+} + 2e^- \rightarrow Sn^{2+}$

$$E^\circ = 0.151 \text{ V} - (-0.762 \text{ V}) = 0.913 \text{ V}$$

$$\Delta_r G^\circ = -\nu F E^\circ = -2 \left(96500 \text{ C mol}^{-1} \right) (0.913 \text{ V}) = -1.762 \times 10^5 \text{ J mol}^{-1} = -1.76 \times 10^5 \text{ J mol}^{-1}$$

$$K = \exp\left(-\frac{\Delta_r G^\circ}{RT} \right) = \exp\left[-\frac{\left(-1.762 \times 10^5 \text{ J mol}^{-1} \right)}{\left(8.314 \text{ J K}^{-1} \text{mol}^{-1} \right) (298.2 \text{ K})} \right] = 7.34 \times 10^{30}$$

(b)   Anode: $2I^- \rightarrow I_2 + 2e^-$

                             Cathode: $Cl_2 + 2e^- \rightarrow 2Cl^-$

$$E^\circ = 1.368 \text{ V} - 0.536 \text{ V} = 0.824 \text{ V}$$

$$\Delta_r G^\circ = -\nu F E^\circ = -2 \left(96500 \text{ C mol}^{-1}\right)(0.824 \text{ V}) = -1.590 \times 10^5 \text{ J mol}^{-1} = -1.59 \times 10^5 \text{ J mol}^{-1}$$

$$K = \exp\left(-\frac{\Delta_r G^\circ}{RT}\right) = \exp\left[-\frac{\left(-1.590 \times 10^5 \text{ J mol}^{-1}\right)}{\left(8.314 \text{ J K}^{-1} \text{ mol}^{-1}\right)(298.2 \text{ K})}\right] = 7.12 \times 10^{27}$$

(c)   Anode: $Fe^{2+} \rightarrow Fe^{3+} + e^-$

                       Cathode: $MnO_4^- + 8H^+ + 5e^- \rightarrow Mn^{2+} + + 4H_2O$

$$E^\circ = 1.507 \text{ V} - 0.771 \text{ V} = 0.736 \text{ V}$$

$$\Delta_r G^\circ = -\nu F E^\circ = -5 \left(96500 \text{ C mol}^{-1}\right)(0.739 \text{ V}) = -3.551 \times 10^5 \text{ J mol}^{-1} = -3.55 \times 10^5 \text{ J mol}^{-1}$$

$$K = \exp\left(-\frac{\Delta_r G^\circ}{RT}\right) = \exp\left[-\frac{\left(-3.551 \times 10^5 \text{ J mol}^{-1}\right)}{\left(8.314 \text{ J K}^{-1} \text{ mol}^{-1}\right)(298.2 \text{ K})}\right] = 1.60 \times 10^{62}$$

7.10   Consider a concentration cell consisting of two hydrogen electrodes. At 25° C, the cell emf is found to be 0.0267 V. If the pressure of hydrogen gas at the anode is 4.0 bar, what is the pressure of hydrogen gas at the cathode?

Let the pressure of hydrogen gas at the cathode be x bars. The half reactions are

                 Anode: $H_2 \ (4.0 \text{ bar}) \rightarrow 2H^+ + 2e^-$

                 Cathode: $2H^+ + 2e^- \rightarrow H_2 \ (x \text{ bar})$

The overall reaction is

$$H_2 \ (4.0 \text{ bar}) \rightarrow H_2 \ (x \text{ bar})$$

Since this is a concentration cell, $E^\circ = 0$ V. The emf for the cell is

$$E = E^\circ - \frac{0.0257 \text{ V}}{\nu} \ln \frac{x}{4.0}$$

$$0.0267 \text{ V} = 0 \text{ V} - \frac{0.0257 \text{ V}}{2} \ln \frac{x}{4.0}$$

$$x = 0.50 \text{ bar}$$

7.12   From the standard reduction potentials listed in Table 7.1 for $Sn^{2+} \mid Sn$ and $Pb^{2+} \mid Pb$, calculate the ratio of $[Sn^{2+}]$ to $[Pb^{2+}]$ at equilibrium at 25° C and the $\Delta_r G^\circ$ value for the reaction.

The reduction potentials for the half cells are

$$Sn^{2+} + 2e^- \rightarrow Sn \qquad E^\circ = -0.138 \text{ V}$$

$$Pb^{2+} + 2e^- \rightarrow Pb \qquad E^\circ = -0.126 \text{ V}$$

The standard emf for the reaction $Sn + Pb^{2+} \rightarrow Sn^{2+} + Pb$ is

$$E^\circ = -0.126 \text{ V} - (-0.138 \text{ V}) = 0.012 \text{ V}$$

The ratio of $[Sn^{2+}]$ to $[Pb^{2+}]$ at equilibrium is directly related to the equilibrium constant, which can be calculated from E°.

$$K = \frac{[Sn^{2+}]}{[Pb^{2+}]} = \exp\left(\frac{\nu F E^\circ}{RT}\right) = \exp\left[\frac{2\,(96500\text{ C mol}^{-1})\,(0.012\text{ V})}{(8.314\text{ J K}^{-1}\text{mol}^{-1})\,(298.2\text{ K})}\right] = 2.55$$

The standard Gibbs energy is

$$\Delta_r G^\circ = -\nu F E^\circ = -2\left(96500\text{ C mol}^{-1}\right)(0.012\text{ V}) = -2.32 \times 10^3\text{ J mol}^{-1}$$

7.14   Calculate the emf of the following concentration cell at 298 K:

$$\text{Mg}(s) \mid \text{Mg}^{2+}(0.24\ M) \parallel \text{Mg}^{2+}(0.53\ M) \mid \text{Mg}(s)$$

$$\text{Anode:}\quad \text{Mg} \rightarrow \text{Mg}^{2+}\,(0.24\ M) + 2e^-$$

$$\text{Cathode:}\quad \text{Mg}^{2+}\,(0.53\ M) + 2e^- \rightarrow \text{Mg}$$

The overall reaction is

$$\text{Mg}^{2+}\,(0.53\ M) \rightarrow \text{Mg}^{2+}\,(0.24\ M)$$

The emf of the cell depends on the concentrations of Mg^{2+} at both the anode and the cathode.

$$E = E^\circ - \frac{0.0257\text{ V}}{\nu}\ \ln\frac{0.24}{0.53} = 0\text{ V} - \frac{0.0257\text{ V}}{2}\ \ln\frac{0.24}{0.53} = 0.010\text{ V}$$

7.16   For the reaction

$$\text{NAD}^+ + \text{H}^+ + 2e^- \rightarrow \text{NADH}$$

$E^{\circ\prime}$ is -0.320 V at 25° C. Calculate the value of E' at pH = 1. Assume that both NAD^+ and NADH are at unimolar concentration.

$$E' = E^{\circ\prime} - \frac{0.0257\text{ V}}{\nu}\ \ln\frac{[\text{NADH}]}{[\text{NAD}]\left([\text{H}^+]/10^{-7}\right)}$$

$$= -0.320 - \frac{0.0257\text{ V}}{2}\ \ln\frac{1}{(1)\left(0.1/10^{-7}\right)}$$

$$= -0.142\text{ V}$$

7.18   Look up the $E^{\circ\prime}$ values in Table 7.2 for the reactions

$$\text{CH}_3\text{CHO} + 2\text{H}^+ + 2e^- \rightarrow \text{C}_2\text{H}_5\text{OH}$$

$$\text{NAD}^+ + \text{H}^+ + 2e^- \rightarrow \text{NADH}$$

Calculate the equilibrium constant for the following reaction at 298 K.

$$\text{CH}_3\text{CHO} + \text{NADH} + \text{H}^+ \rightleftharpoons \text{C}_2\text{H}_5\text{OH} + \text{NAD}^+$$

$$CH_3CHO + 2H^+ + 2e^- \rightarrow C_2H_5OH \qquad E^{o\prime} = -0.197 \text{ V}$$

$$NAD^+ + H^+ + 2e^- \rightarrow NADH \qquad E^{o\prime} = -0.320 \text{ V}$$

The standard emf for the reaction

$$CH_3CHO + NADH + H^+ \rightleftharpoons C_2H_5OH + NAD^+$$

is

$$E^{o\prime} = -0.197 \text{ V} - (-0.320 \text{ V}) = 0.123 \text{ V}$$

The equilibrium constant for this reaction is

$$K' = \exp\left(\frac{\nu F E^{o\prime}}{RT}\right) = \exp\left[\frac{2\left(96500 \text{ C mol}^{-1}\right)(0.123 \text{ V})}{\left(8.314 \text{ J K}^{-1}\text{mol}^{-1}\right)(298 \text{ K})}\right] = 1.45 \times 10^4$$

7.20 Calculate the number of moles of cytochrome c^{3+} formed from cytochrome c^{2+} with the Gibbs energy derived from the oxidation of 1 mole of glucose. ($\Delta_r G^o = -2879$ kJ for the degradation of 1 mole of glucose to CO_2 and H_2O.)

The standard emf for the oxidation of cyt c^{2+} to cyt c^{3+} is

$$Fe^{2+}(\text{cyt } c^{2+}) \rightarrow Fe^{3+}(\text{cyt } c^{3+}) + e^- \qquad E^{o\prime} = -0.254 \text{ V}$$

$\Delta_r G^o$ for the oxidation reaction is the same as $\Delta_r G^{o\prime}$, since no H^+ is involved.

$$\Delta_r G^o = \Delta_r G^{o\prime} = -\nu F E^{o\prime} = -\left(96500 \text{ C mol}^{-1}\right)(-0.254 \text{ V}) = 2.451 \times 10^4 \text{ J mol}^{-1}$$

The number of moles of cyt c^{3+} formed from the oxidation of 1 mole of glucose is

$$\frac{2879 \times 10^3 \text{ J}}{2.451 \times 10^4 \text{ J mol}^{-1}} = 117 \text{ mol}$$

7.22 The oxidation of malate to oxaloacetate is a key reaction in the citric acid cycle:

$$\text{malate} + NAD^+ \rightarrow \text{oxaloacetate} + NADH + H^+$$

Calculate the value of $\Delta_r G^{o\prime}$ and the equilibrium constant for the reaction at pH 7 and 298 K.

The half reactions are

$$\text{Anode:} \quad \text{malate} \rightarrow \text{oxaloacetate} + 2H^+ + 2e^-$$

$$\text{Cathode:} \quad NAD^+ + H^+ + 2e^- \rightarrow NADH$$

The standard emf for the reaction is

$$E^{o\prime} = -0.320 \text{ V} - (-0.166 \text{ V}) = -0.154 \text{ V}$$

The standard Gibbs energy and the equilibrium constant are

$$\Delta_r G^{o\prime} = -\nu F E^{o\prime} = -2 \left(96500 \text{ C mol}^{-1}\right)(-0.154 \text{ V}) = 2.972 \times 10^4 \text{ J mol}^{-1} = 2.97 \times 10^4 \text{ J mol}^{-1}$$

$$K' = \exp\left(-\frac{\Delta_r G^{o\prime}}{RT}\right) = \exp\left[-\frac{2.972 \times 10^4 \text{ J mol}^{-1}}{(8.314 \text{ J K}^{-1}\text{mol}^{-1})(298 \text{ K})}\right] = 6.17 \times 10^{-6}$$

7.24 Flavin adenine dinucleotide (FAD) participates in several biological redox reactions according to the half-reaction

$$\text{FAD} + 2\text{H}^+ + 2e^- \rightarrow \text{FADH}_2$$

If the value of $E^{o\prime}$ of this couple is -0.219 V at 298 K and pH 7, calculate its reduction potential at this temperature and pH when the solution contains **(a)** 85% of the oxidized form and **(b)** 15% of the oxidized form.

At pH = 7 and 298 K, the emf for the reduction of FAD is

$$E' = E^{o\prime} - \frac{0.0257 \text{ V}}{\nu}\ln\frac{[\text{FADH}_2]}{[\text{FAD}]\left([\text{H}^+]/10^{-7}\right)^2}$$

$$= -0.219 \text{ V} - \frac{0.0257 \text{ V}}{2}\ln\frac{[\text{FADH}_2]}{[\text{FAD}]\left(10^{-7}/10^{-7}\right)^2}$$

$$= -0.219 \text{ V} - 0.01285 \text{ V}\ln\frac{[\text{FADH}_2]}{[\text{FAD}]}$$

(a)
$$E' = -0.219 \text{ V} - 0.01285 \text{ V}\ln\frac{0.15}{0.85} = -0.197 \text{ V}$$

(b)
$$E' = -0.219 \text{ V} - 0.01285 \text{ V}\ln\frac{0.85}{0.15} = -0.241 \text{ V}$$

7.26 The nitrite in soil is oxidized to nitrate by the bacteria *nitrobacter agilis* in the presence of oxygen. The half-reduction reactions are

$$\text{NO}_3^- + 2\text{H}^+ + 2e^- \rightarrow \text{NO}_2^- + \text{H}_2\text{O} \qquad E^{o\prime} = 0.42 \text{ V}$$

$$\frac{1}{2}\text{O}_2 + 2\text{H}^+ + 2e^- \rightarrow \text{H}_2\text{O} \qquad E^{o\prime} = 0.82 \text{ V}$$

Calculate the yield of ATP synthesis per mole of nitrite oxidized, assuming an efficiency of 55%. (The $\Delta_r G^{o\prime}$ value for ATP synthesis from ADP and P_i is 31.4 kJ mol^{-1}.)

The standard emf when 1 mole of nitrite is oxidized is

$$E^{o\prime} = 0.82 \text{ V} - 0.42 \text{ V} = 0.40 \text{ V}$$

and the standard Gibbs energy is

$$\Delta_r G^{o\prime} = -nFE^{o\prime} = -2\left(96500 \text{ C mol}^{-1}\right)(0.40 \text{ V}) = -7.72 \times 10^4 \text{ J mol}^{-1}$$

Therefore, the yield of ATP synthesis per mole of nitrite oxidized is

$$\frac{7.72 \times 10^4 \text{ J}}{31.4 \times 10^3 \text{ J (mol ATP)}^{-1}} \times 55\% = 1.4 \text{ mol ATP}$$

7.28 A membrane permeable only to K^+ ions is used to separate the following two solutions:

$$\alpha \quad [\text{KCl}] = 0.10 \text{ } M \quad\quad [\text{NaCl}] = 0.050 \text{ } M$$
$$\beta \quad [\text{KCl}] = 0.050 \text{ } M \quad\quad [\text{NaCl}] = 0.10 \text{ } M$$

Calculate the membrane potential at 25° C, and determine which solution has the more negative potential.

The membrane potential can only be affected by K^+ ions.

$$\Delta E_{K^+} = E_{K^+,\alpha} - E_{K^+,\beta} = \frac{0.0257 \text{ V}}{\nu} \ln \frac{[K^+]_\beta}{[K^+]_\alpha} = 0.0257 \text{ V } \ln \frac{0.050}{0.10}$$

$$= -1.8 \times 10^{-2} \text{ V} = -18 \text{ mV}$$

The α solution has the more negative potential.

7.30 Look up the values of E° for the following half-cell reactions:

$$Ag^+ + e^- \rightarrow Ag$$
$$AgBr + e^- \rightarrow Ag + Br^-$$

Describe how you would use these values to determine the solubility product (K_{sp}) of AgBr at 25° C.

$$Ag^+ + e^- \rightarrow Ag \quad\quad E^\circ = 0.800 \text{ V}$$
$$AgBr + e^- \rightarrow Ag + Br^- \quad\quad E^\circ = 0.0713 \text{ V}$$

The solubility product of AgBr is the equilibrium constant for the following reaction:

$$AgBr \rightarrow Ag^+ + Br^-$$

and the standard emf for this reaction is given by those for the half reactions above.

$$E^\circ = 0.0713 \text{ V} - 0.800 \text{ V} = -0.7287 \text{ V}$$

Therefore,

$$K_{sp} = \exp\left(\frac{\nu F E^\circ}{RT}\right) = \exp\left[\frac{(96500 \text{ C mol}^{-1})\ (-0.7287 \text{ V})}{(8.314 \text{ J K}^{-1} \text{ mol}^{-1})\ (298.2 \text{ K})}\right] = 4.81 \times 10^{-13}$$

7.32 One way to prevent a buried iron pipe from rusting is to connect it with a piece of wire to a magnesium or zinc rod. What is the electrochemical principle for this action?

Both $Mg^{2+} \mid Mg (E^\circ = -2.372$ V) and $Zn^{2+} \mid Zn(E^\circ = -0.762$ V) are more electropositive than $Fe^{2+} \mid Fe(E^\circ = -0.447$ V). The more electropositive metals will be preferentially oxidized, protecting the iron pipe. These protective electrodes are often called sacrificial anodes.

7.34 Given that the $\Delta_r S^\circ$ value for the Daniell cell is -21.7 J K^{-1} mol^{-1}, calculate the temperature coefficient $\left(\partial E^\circ / \partial T\right)_P$ of the cell and the emf of the cell at 80° C.

The temperature coefficient is directly related to $\Delta_r S^\circ$.

$$\Delta_r S^\circ = \nu F \left(\frac{\partial E^\circ}{\partial T}\right)_P$$

$$\left(\frac{\partial E^\circ}{\partial T}\right)_P = \frac{\Delta_r S^\circ}{\nu F} = \frac{-21.7 \text{ J K}^{-1} \text{ mol}^{-1}}{2\left(96500 \text{ C mol}^{-1}\right)}$$

$$= -1.124 \times 10^{-4} \text{ V K}^{-1} = -1.12 \times 10^{-4} \text{ V K}^{-1}$$

The standard emf of the cell at 80° C can be calculated using the temperature coefficient and the standard emf (1.104 V) at 25° C.

$$\left(\frac{\partial E^\circ}{\partial T}\right)_P = \frac{E^\circ_{353.2 \text{ K}} - E^\circ_{298.2 \text{ K}}}{353.2 \text{ K} - 298.2 \text{ K}} = \frac{E^\circ_{353.2 \text{ K}} - 1.104 \text{ V}}{55.0 \text{ K}} = 1.124 \times 10^{-4} \text{ V K}^{-1}$$

$$E^\circ_{353.2 \text{ K}} = 1.098 \text{ V}$$

Note that because of the small temperature coefficient, E° changes only slightly with temperature.

7.36 Given the following standard reduction potentials, calculate the ion-product K_w value $\left(\left[H^+\right]\left[OH^-\right]\right)$ at 25° C:

$$2H^+(aq) + 2e^- \rightarrow H_2(g) \qquad E^\circ = 0.00 \text{ V}$$

$$2H_2O(l) + 2e^- \rightarrow H_2(g) + 2OH^-(aq) \qquad E^\circ = -0.828 \text{ V}$$

The ion-product is the equilibrium constant for the following process.

$$H_2O \rightleftharpoons H^+ + OH^-$$

The standard emf for this reaction can be calculated using those for the half-cells described in the question.

$$E^\circ = -0.828 \text{ V} - 0 \text{ V} = -0.828 \text{ V}$$

The ion-product is calculated from E°.

$$K_w = \exp\left(\frac{\nu F E^\circ}{RT}\right) = \exp\left[\frac{\left(96500 \text{ C mol}^{-1}\right)\left(-0.828 \text{ V}\right)}{\left(8.314 \text{ J K}^{-1} \text{ mol}^{-1}\right)\left(298.2 \text{ K}\right)}\right] = 1.01 \times 10^{-14}$$

7.38   The magnitudes of the standard electrode potentials of two metals, X and Y, are

$$X^{2+} + 2e^- \rightarrow X \left| E^\circ \right| = 0.25 \text{ V}$$

$$Y^{2+} + 2e^- \rightarrow Y \left| E^\circ \right| = 0.34 \text{ V}$$

where the | | notation denotes that only the magnitude (but *not* the sign) of the E° value is shown. When the half-cells of X and Y are connected, electrons flow from X to Y. When X is connected to a SHE, electrons flow from X to SHE. **(a)** Which value of E° is positive and which is negative? **(b)** What is the standard emf of a cell made up of X and Y?

(a) Since electrons flow from X to SHE, the anode reaction involves X^{2+} | X. The standard emf for a SHE is 0 V. Therefore,

$$E^\circ = E^\circ_{\text{cathode}} - E^\circ_{\text{anode}} = 0 - E^\circ_{\text{anode}}$$

must be positive. In other words, the standard emf for

$$X^{2+} + 2e^- \rightarrow X$$

is -0.25 V.

Since electrons flow from X to Y, the anode reaction involves X^{2+} | X and the cathode reaction involves Y^{2+} | Y:

$$\text{Anode:} \quad X \rightarrow X^{2+} + 2e^- \quad E^\circ = 0.25 \text{ V}$$
$$\text{Cathode:} \quad Y^{2+} + 2e^- \rightarrow Y \quad | E^\circ |= 0.34 \text{ V}$$

The emf for the reaction must be positive. Thus, the standard reduction potential for the Y^{2+} | Y pair is positive, that is, $E^\circ = 0.34$ V, and the standard reduction potential for the X^{2+} | X pair is negative, that is, $E^\circ = -0.25$ V.

(b) The standard emf of a cell made up of X and Y is

$$E^\circ = 0.34 \text{ V} - (-0.25 \text{ V}) = 0.59 \text{ V}$$

7.40   Given the standard reduction potential for Au^{3+} in Table 7.1 and

$$Au^+(aq) + e^- \rightarrow Au(s) \qquad E^\circ = 1.69 \text{ V}$$

answer the following questions. **(a)** Why does gold not tarnish in air? **(b)** Will the following disproportionation occur spontaneously?

$$3Au^+(aq) \rightarrow Au^{3+}(aq) + 2Au(s)$$

(c) Predict the reaction between gold and fluorine gas.

(a) Gold does not tarnish (oxidize) in air because the reduction potential for oxygen is insufficient to oxidize gold (either to Au^+ as given or to Au^{3+} with $E^\circ = 1.498$ V).

$$O_2 + 4H^+ + 4e^- \rightarrow 2H_2O \qquad E^\circ = 1.229 \text{ V}$$

(b) The disproportionation can be considered as the sum of the two half reactions,

$$3\left(Au^+ + e^- \rightarrow Au\right)$$

$$Au \rightarrow Au^{3+} + 3e^-$$

so that the standard emf for the reaction is

$$E^\circ = 1.692\ V - 1.498\ V = 0.194\ V.$$

Since E° for the reaction is positive, the disproportionation will be spontaneous under standard state conditions.

(c) With $E^\circ = 2.87\ V$, the half reaction

$$F_2 + 2e^- \rightarrow 2F^-$$

is able to oxidize gold completely to Au^{3+} via

$$3F_2 + 2Au \rightarrow 2AuF_3$$

7.42 Calculate the pressure of H_2 (in bar) required to maintain equilibrium with respect to the following reaction at $25°\ C$

$$Pb(s) + 2H^+(aq) \rightleftharpoons Pb^{2+}(aq) + H_2(g)$$

given that $[Pb^{2+}] = 0.035\ M$ and the solution is buffered at pH 1.60.

The cell reaction is

$$\text{Anode:}\qquad Pb \rightarrow Pb^{2+} + 2e^-$$
$$\text{Cathode:}\qquad 2H^+ + 2e^- \rightarrow H_2$$

The standard emf for the reaction

$$Pb(s) + 2H^+(aq) \rightleftharpoons Pb^{2+}(aq) + H_2(g)$$

is

$$E^\circ = 0\ V - (-0.126\ V) = 0.126\ V$$

The emf of the cell is

$$E = E^\circ - \frac{0.0257\ V}{\nu}\ \ln \frac{\left[Pb^{2+}\right] P_{H_2}}{\left[H^+\right]^2}$$

From the pH, $[H^+] = 10^{-1.60} = 2.51 \times 10^{-2}\ M$. At equilibrium, $E = 0$. Therefore, the above equation becomes

$$0 = 0.126 - \frac{0.0257\ V}{2}\ \ln \frac{(0.035)\ P_{H_2}}{\left(2.51 \times 10^{-2}\right)^2}$$

$$P_{H_2} = 3.3 \times 10^2\ \text{bar}$$

7.44   Use the data in Table 7.1 to determine the value of $\Delta_f \overline{G}^{\,\circ}$ for $Fe^{2+}(aq)$.

The reduction potential for the $Fe^{2+} \mid Fe$ couple is

$$Fe^{2+} + 2e^- \rightarrow Fe \qquad E^\circ = -0.447 \text{ V}$$

Consider a cell constructed using the following half-cells:

$$\text{Anode:} \qquad Fe(s) \rightarrow Fe^{2+}(aq) + 2e^-$$

$$\text{Cathode:} \qquad 2H^+(aq) + 2e^- \rightarrow H_2(g)$$

The overall equation and standard emf are

$$Fe(s) + 2H^+(aq) \rightarrow Fe^{2+}(aq) + H_2(g) \qquad E^\circ = 0.447 \text{ V}$$

The standard Gibbs energy for this reaction is related to both the standard Gibbs energy of formation of the reactants and products and the standard emf of the reaction.

$$\Delta_r G^\circ = -\nu F E^\circ = \Delta_f \overline{G}^{\,\circ}\left[Fe^{2+}(aq)\right] + \Delta_f \overline{G}^{\,\circ}\left[H_2(g)\right] - \Delta_f \overline{G}^{\,\circ}[Fe(s)] - 2\Delta_f \overline{G}^{\,\circ}\left[H^+(aq)\right]$$

$$-2\left(96500 \text{ C mol}^{-1}\right)(0.447 \text{ V}) = \Delta_f \overline{G}^{\,\circ}\left[Fe^{2+}(aq)\right] + 0 \text{ kJ mol}^{-1} - 0 \text{ kJ mol}^{-1} - 2\left(0 \text{ kJ mol}^{-1}\right)$$

$$\Delta_f \overline{G}^{\,\circ}\left[Fe^{2+}(aq)\right] = -8.63 \times 10^4 \text{ J mol}^{-1}$$

7.46   Calculate the E° for the propane fuel cell discussed on p. 246. The $\Delta_f \overline{G}^{\,\circ}$ for C_3H_8 is $-23.5 \text{ kJ mol}^{-1}$.

The half reactions and overall reaction for the fuel cell are

$$\text{Anode:} \qquad C_3H_8(g) + 6H_2O(l) \rightarrow 3CO_2(g) + 20H^+(aq) + 20e^-$$

$$\text{Cathode:} \qquad 5O_2(g) + 20H^+(aq) + 20e^- \rightarrow 10H_2O(l)$$

$$\text{Overall:} \qquad C_3H_8(g) + 5O_2(g) \rightarrow 3CO_2(g) + 4H_2O(l)$$

$\Delta_r G^\circ$ for the overall reaction is

$$\Delta_r G^\circ = 3\Delta_f \overline{G}^{\,\circ}\left[CO_2(g)\right] + 4\Delta_f \overline{G}^{\,\circ}\left[H_2(l)\right] - \Delta_f \overline{G}^{\,\circ}\left[C_3H_8(g)\right] - 5\Delta_f \overline{G}^{\,\circ}\left[O_2(g)\right]$$

$$= 3\left(-394.4 \text{ kJ mol}^{-1}\right) + 4\left(-237.2 \text{ kJ mol}^{-1}\right) - \left(-23.5 \text{ kJ mol}^{-1}\right) - 5\left(0 \text{ kJ mol}^{-1}\right)$$

$$= -2108.5 \text{ kJ mol}^{-1}$$

E° can now be calculated from $\Delta_r G^\circ$ using the relation $\Delta_r G^\circ = -\nu F E^\circ$. (Note that $\nu = 20$, as indicated by the half reactions.)

$$E^\circ = -\frac{\Delta_r G^\circ}{\nu F} = -\frac{-2108.5 \times 10^3 \text{ J mol}^{-1}}{20\left(96500 \text{ C mol}^{-1}\right)} = 1.09 \text{ V}$$

CHAPTER 8

Acids and Bases

PROBLEMS AND SOLUTIONS

8.2 Write the formulas for the conjugate bases of the following acids: **(a)** HI, **(b)** H_2SO_4, **(c)** H_2S, **(d)** HCN, and **(e)** HCOOH (formic acid).

(a) I^-

(b) HSO_4^-

(c) HS^-

(d) CN^-

(e) $HCOO^-$

8.4 Classify each of the following species as a weak or strong base: **(a)** LiOH, **(b)** CN^-, **(c)** H_2O, **(d)** ClO_4^-, and **(e)** NH_2^-.

(a) Actually LiOH is not a Brønsted base, but it dissociates into the ions Li^+ and OH^- in aqueous solution. OH^- is a strong base (the strongest base that can exist in aqueous solution).

(b) Weak,

(c) weak,

(d) weak,

(e) strong.

8.6 A 0.040 M solution of a monoprotic acid is 13.5% dissociated. What is the dissociation constant of the acid?

The dissociation of the acid can be represented by

$$HA \rightleftharpoons H^+ + A^-$$

At equilibrium,

$$[H^+] = [A^-] = (0.135)\,(0.040\,M) = 5.40 \times 10^{-3}\,M$$

$$[HA] = (1 - 0.135)\,(0.040\,M) = 3.46 \times 10^{-2}\,M$$

The dissociation constant, K_a, of the acid is therefore

$$K_a = \frac{[H^+]\,[A^-]}{[HA]} = \frac{(5.40 \times 10^{-3})^2}{3.46 \times 10^{-2}} = 8.4 \times 10^{-4}$$

8.8   The dissociation constant of a monoprotic acid at 298 K is 1.47×10^{-3}. Calculate the degree of dissociation by **(a)** assuming ideal behavior and **(b)** using a mean activity coefficient $\gamma_\pm = 0.93$. The concentration of the acid is $0.010\,M$.

Let α be the degree of dissociation of the monoprotic acid. The corresponding concentrations of all species are

	HA	$\rightleftharpoons$	H^+	+	A^-	
Initial	0.010		0		0	M
At equilibrium	$0.010\,(1 - \alpha)$		0.010α		0.010α	M

(a)

$$K_a = 1.47 \times 10^{-3} = \frac{[H^+]\,[A^-]}{[HA]} = \frac{(0.010\alpha)^2}{0.010\,(1 - \alpha)}$$

$$1.0 \times 10^{-4}\alpha^2 + 1.47 \times 10^{-5}\alpha - 1.47 \times 10^{-5} = 0$$

$$\alpha = 0.32$$

Therefore, assuming ideal behavior, the acid is 32% dissociated.

(b)

$$K_a = \frac{a_{H^+}a_{A^-}}{a_{HA}} = \frac{[H^+]\,\gamma_+\,[A^-]\,\gamma_-}{[HA]\,\gamma_{HA}}$$

Since HA is an uncharged species and the solution is dilute, γ_{HA} is approximately 1. Furthermore, $\gamma_+\gamma_- = \gamma_\pm^2$. The K_a expression becomes

$$K_a = 1.47 \times 10^{-3} = \frac{[H^+]\,[A^-]\,\gamma_\pm^2}{[HA]} = \frac{(0.010\alpha)^2\,(0.93)^2}{0.010\,(1 - \alpha)}$$

$$8.65 \times 10^{-5}\alpha^2 + 1.47 \times 10^{-5}\alpha - 1.47 \times 10^{-5} = 0$$

$$\alpha = 0.34$$

Therefore, accounting for non-ideality, the acid is 34% dissociated.

8.10   HF is a weak acid, but its strength increases with concentration. Explain. (*Hint*: F^- reacts with HF to form HF_2^-. The equilibrium constant for this reaction is 5.2 at 25° C.)

In addition to the dissociation of HF,

$$HF \rightleftharpoons H^+ + F^-$$

there is also the reaction

$$F^- + HF \rightleftharpoons HF_2^-$$

The second equilibrium has a fairly large equilibrium constant ($K = 5.2$) so that much of the F^- produced in the first reaction is removed in the second step. Applying Le Chatelier's principle, more HF must dissociate to compensate for the removal of F^-, at the same time producing more H^+. With increasing HF concentration, Le Chatelier's principle applied to the second equation requires that even more F^- be removed and thus further dissociation of HF.

8.12 What are the concentrations of HSO_4^-, SO_4^{2-}, and H^+ in a 0.20 M $KHSO_4$ solution? (*Hint:* H_2SO_4 is a strong acid; K_a for $HSO_4^- = 1.3 \times 10^{-2}$.)

KHSO$_4$ is a strong electrolyte and dissociates completely into K^+ and HSO_4^-. K^+ is neither a Brønsted acid nor a Brønsted base, and it does not hydrolyze. HSO_4^- functions as a Brønsted acid and a Brønsted base. As the Brønsted base of a strong acid, H_2SO_4, HSO_4^- does not hydrolyze to any significant extent to give the acid. As a Brønsted acid, it dissociates and makes the solution acidic. Assume that x M of HSO_4^- dissociates in the solution, then the equilibrium concentrations of various species are

$$
\begin{array}{cccc}
HSO_4^- & \rightleftharpoons & H^+ + & SO_4^{2-} \\
\end{array}
$$

At equilibrium $\quad 0.20 - x \qquad\quad x \qquad\quad x \quad M$

x is determined using the equilibrium expression

$$K_a = 1.3 \times 10^{-2} = \frac{[H^+]\left[SO_4^{2-}\right]}{[HSO_4^-]} = \frac{x^2}{0.2 - x}$$

$$x^2 + 1.3 \times 10^{-2}x - 2.6 \times 10^{-3} = 0$$

$$x = 4.49 \times 10^{-2}$$

Therefore,

$$\left[HSO_4^-\right] = (0.20 - x) \ M = 0.16 \ M$$

$$\left[H^+\right] = x \ M = 4.5 \times 10^{-2} \ M$$

$$\left[SO_4^{2-}\right] = x \ M = 4.5 \times 10^{-2} \ M$$

8.14 To which of the following would the addition of an equal volume of 0.60 M NaOH lead to a solution having a lower pH? (**a**) Water, (**b**) 0.30 M HCl, (**c**) 0.70 M KOH, and (**d**) 0.40 M NaNO$_3$.

(**a**) The original solution has pH = 7.00, assuming it to be pure water. After the addition of an equal volume of 0.60 M NaOH, the solution is 0.30 M NaOH with pH = 13.48, which is higher.

(**b**) The original solution has pH = −0.52. After the addition of an equal volume of 0.60 M NaOH, the solution is 0.15 M NaOH (plus NaCl) with pH = 13.18, which is higher.

(**c**) The original solution has pH = 13.85. After the addition of an equal volume of 0.60 M NaOH, the solution is 0.65 M in OH$^-$ with pH = 13.81, which is lower.

(d) Since neither ion hydrolyzes to any extent, the original pH = 7.00, and the final pH is the same as in part **(a)**, namely 13.48, which is higher.

Only the addition to solution **(c)** leads to a solution with lower pH.

8.16   A solution of methylamine (CH_3NH_2) has a pH of 10.64. How many grams of methylamine are in 100.0 mL of the solution?

Methylamine is a base. It ionizes to give equal concentrations of $CH_3NH_3^+$ and OH^-, which can be calculated from the pH of the solution.

$$pOH = 14.00 - pH = 14.00 - 10.64 = 3.36$$

$$[OH^-] = 10^{-3.36} = 4.37 \times 10^{-4}\ M$$

Let $x\ M$ of methylamine be present initially. The equilibrium concentration is then $x - 4.37 \times 10^{-4}\ M$. The concentrations of various species are written below the chemical equation.

	CH_3NH_2	+	H_2O	⇌	$CH_3NH_3^+$	+	OH^-	
Initial	x				0		0	M
At equilibrium	$x - 4.37 \times 10^{-4}$				4.37×10^{-4}		4.37×10^{-4}	M

$$K_b = 4.38 \times 10^{-4} = \frac{[CH_3NH_3^+][OH^-]}{[CH_3NH_2]} = \frac{(4.37 \times 10^{-4})^2}{x - 4.37 \times 10^{-4}}$$

$$x = 8.73 \times 10^{-4}$$

where x represents the initial concentration of CH_3NH_2 in solution.

The mass of methylamine can now be calculated.

$$\text{Number of moles of methylamine} = \left(8.73 \times 10^{-4}\ M\right)(0.100\ L) = 8.73 \times 10^{-5}\ mol$$

$$\text{Mass of methylamine} = \left(8.73 \times 10^{-5}\ mol\right)\left(31.06\ g\ mol^{-1}\right) = 2.7 \times 10^{-3}\ g$$

8.18   Novocaine, used as a local anesthetic by dentists, is a weak base ($K_b = 8.91 \times 10^{-6}$). What is the ratio of the concentration of the base to that of its acid in the blood plasma (pH = 7.40) of a patient?

The equilibrium between novocaine, NOV, and its acid, $HNOV^+$, can be expressed as

$$NOV + H_2O \rightleftharpoons HNOV^+ + OH^-$$

The concentration of OH^- is related to the pH of the blood plasma.

$$pOH = 14.00 - pH = 14.00 - 7.40 = 6.60$$

$$[OH^-] = 10^{-6.60} = 2.51 \times 10^{-7}\ M$$

The equilibrium expression involving novocaine and its acid is

$$K_b = \frac{[HNOV^+][OH^-]}{[NOV]}$$

Therefore,

$$\frac{[NOV]}{[HNOV^+]} = \frac{[OH^-]}{K_b} = \frac{2.51 \times 10^{-7}}{8.91 \times 10^{-6}} = 2.8 \times 10^{-2}$$

8.20 Explain why phenol is a stronger acid than methanol:

phenol

CH_3-OH

methanol

The two conjugate bases are $C_6H_5O^-$ from phenol and CH_3O^- from methanol. The $C_6H_5O^-$ is stabilized by resonance:

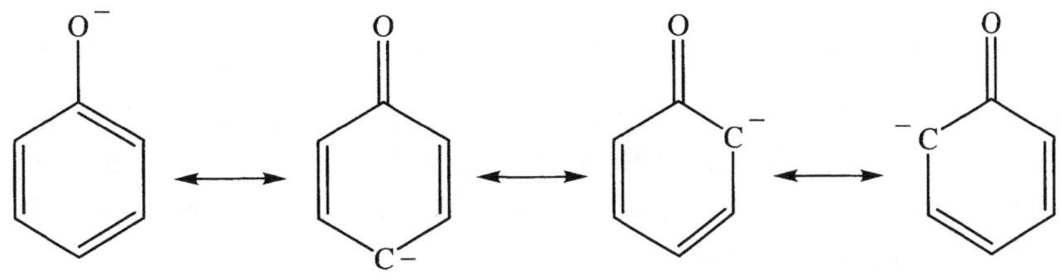

There is no such resonance stabilization for CH_3O^-. A more stable conjugate base means an increase in the strength of the acid.

8.22 The disagreeable odor of fish is mainly due to organic compounds (RNH_2) containing an amino group, $-NH_2$, where R is the rest of the molecule. Amines are bases just like ammonia. Explain why putting some lemon juice on fish can greatly reduce the odor.

Although primary amines ($R-NH_2$) interact through dipole-dipole attractions and may also participate in hydrogen bonding, they are relatively volatile. When protonated to form a salt, $R-NH_3^+$, the formal electrostatic intermolecular attractions result in much lower vapor pressures and consequently less odor. Lemon juice contains acids which react with the primary amines to form ammonium salts.

8.24 Calculate the pH of a 0.10 M NH$_4$Cl solution.

NH_4^+ hydrolyzes in water as described by the following equation:

$$NH_4^+ \quad + \quad H_2O \quad \rightleftharpoons \quad NH_3 \quad + \quad H_3O^+$$

	NH_4^+			NH_3	H_3O^+	
Initial	0.10			0	0	M
At equilibrium	0.10 − x			x	x	M

$$K_a = \frac{K_w}{K_b\,(NH_3)} = \frac{1.0 \times 10^{-14}}{1.8 \times 10^{-5}} = \frac{[NH_3]\,[H_3O^+]}{[NH_4^+]}$$

$$5.56 \times 10^{-10} = \frac{x^2}{0.10 - x}$$

Since K_a is very small, $0.10 - x$ can be approximated as 0.10, and the expression above simplifies to

$$5.56 \times 10^{-10} = \frac{x^2}{0.10}$$

$$x = 7.46 \times 10^{-6}$$

Checking the assumption,

$$\frac{x}{0.10} \times 100\% = 7.5 \times 10^{-3}\% < 5\%$$

Thus, the assumption is valid. The pH of the solution is

$$pH = -\log\left(7.46 \times 10^{-6}\right) = 5.13$$

8.26   A student added NaOH solution from a buret to an Erlenmeyer flask containing HCl solution and used phenolphthalein as indicator. At the equivalence point of the titration, she observed a faint reddish-pink color. However, after a few minutes, the solution gradually turned colorless. What do you suppose happened?

CO_2 in the air is absorbed by the solution where it is converted to carbonic acid,

$$CO_2 + H_2O \rightleftharpoons H_2CO_3$$

The carbonic acid neutralizes the excess NaOH, lowering the pH sufficiently to render the phenolphthalein colorless.

8.28   The K_a of a certain indicator is 2.0×10^{-6}. The color of HIn is green, and that of In$^-$ is red. A few drops of the indicator are added to a HCl solution, which is then titrated against a NaOH solution. At what pH will the indicator change color?

When [HIn] = [In$^-$], the indicator color is a mixture of the colors of HIn and In$^-$. In other words, the indicator changes color at this point, and the pH can be calculated using the Henderson–Hasselbalch equation:

$$pH = pK_a + \log\frac{[In^-]}{[HIn]} = -\log\left(2.0 \times 10^{-6}\right) = 5.70$$

8.30   A 200-mL volume of NaOH solution was added to 400 mL of a 2.00 M HNO_2 solution. The pH of the mixed solution was 1.50 units greater than that of the original acid solution. Calculate the molarity of the NaOH solution.

First calculate the pH of the original solution.

$$HNO_2 \rightleftharpoons H^+ + NO_2^-$$

Initial	2.00	0	0 M
At equilibrium	2.00 − x	x	x M

$$K_a = \frac{[H^+][NO_2^-]}{[HNO_2]}$$

$$4.5 \times 10^{-4} = \frac{x^2}{2.00 - x}$$

$$x^2 + 4.5 \times 10^{-4}x - 9.0 \times 10^{-4} = 0$$

$$x = 2.98 \times 10^{-2}$$

Therefore, the pH of the original solution is $- \log x = 1.526$. The pH of the mixed solution is $1.526 + 1.50 = 3.026$.

Let the molarity of the NaOH solution be y M. The concentrations of NaOH and HNO_2 in the mixture before the neutralization reaction are

$$[NaOH] = \frac{(y\ M)(200\ mL)}{600\ mL} = 0.3333y\ M$$

$$[HNO_2] = \frac{(2.00\ M)(400\ mL)}{600\ mL} = 1.333\ M$$

NaOH and HNO_2 react essentially to completion:

$$OH^- + HNO_2 \rightarrow H_2O + NO_2^-$$

After the reaction,

$$[HNO_2] = (1.333 - 0.3333y)\ M$$

$$[NO_2^-] = 0.3333y\ M$$

The value of y can be obtained using the Henderson–Hasselbalch equation:

$$pH = pK_a + \log \frac{[NO_2^-]}{[HNO_2]}$$

$$3.026 = 3.35 + \log \frac{0.3333y}{1.333 - 0.3333y}$$

$$\frac{0.3333y}{1.333 - 0.3333y} = 10^{-0.324} = 0.474$$

$$y = 1.3\ M$$

Therefore, the concentration of the NaOH solution is 1.3 M.

8.32   Phenolphthalein is the common indicator for the titration of a strong acid with a strong base. **(a)** If the pK_a of phenolphthalein is 9.10, what is the ratio of the nonionized form of the indicator (colorless) to the ionized form (reddish pink) at pH 8.00? **(b)** If 2 drops of 0.060 M phenolphthalein are used in a titration involving a 50.0-mL volume, what is the concentration of the ionized form at pH 8.00? (Assume that 1 drop = 0.050 mL.)

(a) The ratio of the nonionized form of phenolphthalein to the ionized form can be calculated using the Henderson–Hasselbalch equation.

$$pH = pK_a + \log \frac{[\text{conjugate base}]}{[\text{acid}]}$$

$$8.00 = 9.10 + \log \frac{[\text{ionized}]}{[\text{nonionized}]}$$

$$\log \frac{[\text{nonionized}]}{[\text{ionized}]} = 1.10$$

$$\frac{[\text{nonionized}]}{[\text{ionized}]} = 12.59 = 12.6$$

(b) The total concentration of the indicator is the sum of the total concentrations of the nonionized and ionized forms.

$$[\text{indicator}] = [\text{nonionized}] + [\text{ionized}] = \frac{(2 \text{ drops})\left(\frac{0.050 \text{ mL}}{1 \text{ drop}}\right)(0.060 \text{ } M)}{50.0 \text{ mL}} = 1.2 \times 10^{-4} \text{ } M$$

The concentration of the ionized form of the indicator can be found using the result of part **(a)**.

$$\frac{[\text{nonionized}]}{[\text{ionized}]} = 12.6 = \frac{[\text{indicator}] - [\text{ionized}]}{[\text{ionized}]} = \frac{1.2 \times 10^{-4} \text{ } M - [\text{ionized}]}{[\text{ionized}]}$$

$$13.6 \, [\text{ionized}] = 1.2 \times 10^{-4} \text{ } M$$

$$[\text{ionized}] = 8.8 \times 10^{-6} \text{ } M$$

8.34   Specify which of the following systems can be classified as a buffer system: **(a)** KCl/HCl, **(b)** NH_3/NH_4NO_3, **(c)** Na_2HPO_4/NaH_2PO_4, **(d)** KNO_2/HNO_2, **(e)** $KHSO_4$/H_2SO_4, and **(f)** HCOOK/HCOOH.

A buffer is composed of a weak acid and a weak base. **(b)**, **(c)**, **(d)**, and **(f)** can be classified as buffer systems.

8.36   Calculate the pH of the 0.20 M NH_3/0.20 M NH_4Cl buffer. What is the pH of the buffer after the addition of 10.0 mL of 0.10 M HCl to 65.0 mL of the buffer?

The pH of the buffer can be calculated using the Henderson–Hasselbalch equation.

$$pH = pK_a + \log\left(\frac{[NH_3]}{[NH_4^+]}\right)$$

$$= 9.25 + \log\left(\frac{0.20}{0.20}\right) = 9.25$$

When HCl is added, it will react with NH_3 to near completion.

$$\text{Initial [HCl] in the reaction mixture} = \frac{(0.10\,M)\,(10.0\,\text{mL})}{75.0\,\text{mL}} = 1.33 \times 10^{-2}\,M$$

$$\text{Initial [NH}_3\text{] in the reaction mixture} = \frac{(0.20\,M)\,(65.0\,\text{mL})}{75.0\,\text{mL}} = 0.173\,M$$

$$\text{Initial [NH}_4^+\text{] in the reaction mixture} = \frac{(0.20\,M)\,(65.0\,\text{mL})}{75.0\,\text{mL}} = 0.173\,M$$

	H^+	+	NH_3	$\rightleftharpoons$	NH_4^+	
Before reaction	1.33×10^{-2}		0.173		0.173	M
After reaction	0		0.160		0.186	M

The new pH can now be calculated using the Henderson–Hasselbalch equation.

$$pH = pK_a + \log\left(\frac{[NH_3]}{[NH_4^+]}\right)$$

$$= 9.25 + \log\left(\frac{0.160}{0.186}\right) = 9.18$$

8.38 A quantity of 26.4 mL of a 0.45 M acetic acid solution is added to 31.9 mL of a 0.37 M sodium hydroxide solution. What is the pH of the final solution?

$$\text{Number of moles of NaOH added} = (0.37\,M)\left(31.9 \times 10^{-3}\,\text{L}\right) = 1.18 \times 10^{-2}\,\text{mol}$$

$$\text{Number of moles of CH}_3\text{COOH initially} = (0.45\,M)\left(26.4 \times 10^{-3}\,\text{L}\right) = 1.19 \times 10^{-2}\,\text{mol}$$

Since OH^- reacts with CH_3COOH in an equimolar manner, nearly all the CH_3COOH will be converted to CH_3COO^- which has a concentration of

$$[CH_3COO^-] = \frac{1.18 \times 10^{-2}\,\text{mol}}{26.4 \times 10^{-3}\,\text{L} + 31.9 \times 10^{-3}\,\text{L}} = 0.202\,M$$

The concentration of the remaining CH_3COOH is

$$[CH_3COOH] = \frac{0.01 \times 10^{-2}\,\text{mol}}{26.4 \times 10^{-3}\,\text{L} + 31.9 \times 10^{-3}\,\text{L}}$$

$$= 2 \times 10^{-3}\,M$$

The equilibrium between CH_3COO^- and CH_3COOH is

	CH_3COO^-	+	H_2O	$\rightleftharpoons$	CH_3COOH	+	OH^-	
Initial	0.202				2×10^{-3}		0	M
At equilibrium	$0.202 - x$				$2 \times 10^{-3} + x$		x	M

$$K_b = \frac{K_w}{K_a\,(CH_3COOH)} = \frac{1.0 \times 10^{-14}}{1.75 \times 10^{-5}} = \frac{[CH_3COOH]\,[OH^-]}{[CH_3COO^-]}$$

$$5.714 \times 10^{-10} = \frac{x\,(2 \times 10^{-3} + x)}{0.202 - x}$$

Since K_b is very small, $0.202 - x$ can be approximated by 0.202, and $2 \times 10^{-3} + x$ by 2×10^{-3}. The expression above simplifies to

$$5.714 \times 10^{-10} = \frac{x\,(2 \times 10^{-3})}{0.202}$$

$$x = 6 \times 10^{-8}$$

Checking the assumption,

$$\frac{x}{0.202} \times 100\% = 3 \times 10^{-5}\% < 5\%$$

$$\frac{x}{2 \times 10^{-3}} \times 100\% = 3 \times 10^{-3}\% < 5\%$$

Therefore, the assumptions are valid. The pH of the solution is related to $[OH^-]$:

$$pOH = -\log 6 \times 10^{-8} = 7.2$$

$$pH = 14.0 - 7.2 = 6.8$$

8.40 A phosphate buffer has a pH equal to 7.30. **(a)** What is the predominant conjugate pair present in this buffer? **(b)** If the concentration of this buffer is $0.10\,M$, what is the new pH after the addition of 5.0 mL of $0.10\,M$ HCl to 20.0 mL of this buffer solution?

(a) The predominant conjugate pair will involve an acid with a pK_a value similar to the pH of the buffer. Since for $H_2PO_4^-$, $pK_a'' = 7.21$, $H_2PO_4^-$ and HPO_4^{2-} are the chief components of this buffer.

(b) First the concentrations of $H_2PO_4^-$ and HPO_4^{2-} need to be calculated. The sum of the concentrations is $0.10\,M$. In other words,

$$\left[HPO_4^{2-}\right] = 0.10\,M - \left[H_2PO_4^-\right]$$

Using the Henderson–Hasselbalch equation,

$$pH = pK_a'' + \log \frac{\left[HPO_4^{2-}\right]}{\left[H_2PO_4^-\right]}$$

$$7.30 = 7.21 + \log \frac{0.10\,M - \left[H_2PO_4^-\right]}{\left[H_2PO_4^-\right]}$$

$$\frac{0.10\,M - \left[H_2PO_4^-\right]}{\left[H_2PO_4^-\right]} = 10^{0.09} = 1.2$$

$$\left[H_2PO_4^-\right] = 4.5 \times 10^{-2}\,M$$

$$\left[HPO_4^{2-}\right] = 0.10\,M - 4.5 \times 10^{-2}\,M = 5.5 \times 10^{-2}\,M$$

Before the addition of HCl,

$$\text{Number of moles of } H_2PO_4^- = \left(4.5 \times 10^{-2}\, M\right)\left(20.0 \times 10^{-3}\, L\right) = 9.0 \times 10^{-4}\, \text{mol}$$

$$\text{Number of moles of } HPO_4^{2-} = \left(5.5 \times 10^{-2}\, M\right)\left(20.0 \times 10^{-3}\, L\right) = 1.1 \times 10^{-3}\, \text{mol}$$

When HCl is added, it will react with HPO_4^{2-}:

$$H^+ + HPO_4^{2-} \rightleftharpoons H_2PO_4^-$$

and the reaction will proceed to near completion.

$$\text{Moles of HCl added} = (0.10\, M)\left(5.0 \times 10^{-3}\, L\right) = 5.0 \times 10^{-4}\, \text{mol}$$

$$\text{Moles of } HPO_4^{2-} \text{ after reaction with HCl} = 1.1 \times 10^{-3}\, \text{mol} - 5.0 \times 10^{-4}\, \text{mol} = 6 \times 10^{-4}\, \text{mol}$$

$$\left[HPO_4^{2-}\right] \text{ after reaction with HCl} = \frac{6 \times 10^{-4}\, \text{mol}}{25.0 \times 10^{-3}\, L} = 0.02\, M$$

$$\text{Moles of } H_2PO_4^- \text{ after reaction with HCl} = 9.0 \times 10^{-4}\, \text{mol} + 5.0 \times 10^{-4}\, \text{mol} = 1.40 \times 10^{-3}\, \text{mol}$$

$$\left[H_2PO_4^{2-}\right] \text{ after reaction with HCl} = \frac{1.40 \times 10^{-3}\, \text{mol}}{25.0 \times 10^{-3}\, L} = 0.056\, M$$

The new pH can now be calculated using the Henderson–Hasselbalch equation.

$$pH = pK_a'' + \log \frac{\left[HPO_4^{2-}\right]}{\left[H_2PO_4^-\right]}$$

$$= 7.21 + \log \frac{0.02}{0.056}$$

$$= 6.76$$

8.42 Describe the number of different ways to prepare 1 liter of a 0.050 M phosphate buffer with a pH of 7.8.

At pH = 7.8, the appropriate buffer couple is $H_2PO_4^-/HPO_4^{2-}$. There are a number of ways to prepare the buffer, but each will require the proper ratio of the two phosphate species as given by the Henderson–Hasselbalch equation,

$$pH = pK_a'' + \log \frac{\left[HPO_4^{2-}\right]}{\left[H_2PO_4^-\right]}$$

$$7.8 = 7.21 + \log \frac{\left[HPO_4^{2-}\right]}{\left[H_2PO_4^-\right]}$$

$$\frac{\left[HPO_4^{2-}\right]}{\left[H_2PO_4^-\right]} = 3.9 \approx 4$$

Thus each liter of solution would require $\frac{4}{5} \times 0.050$ mol $= 0.040$ mol of HPO_4^{2-} and $\frac{1}{5} \times 0.050$ mol $= 0.010$ mol of $H_2PO_4^-$.

The buffer could be prepared as follows. (In each case the K^+ and Na^+ counter ions are interchangeable.)

1. Mix KH_2PO_4 and K_2HPO_4 in the proper proportion
2. Titrate H_3PO_4 with NaOH to obtain the appropriate amounts of NaH_2PO_4 and Na_2HPO_4.
3. Start with NaH_2PO_4 and add NaOH to convert the proper amount to Na_2HPO_4.
4. Start with Na_2HPO_4 and add HCl to convert the appropriate amount to NaH_2PO_4.
5. Start with K_3PO_4 and convert to the correct ratio of KH_2PO_4 and K_2HPO_4 by adding HCl.
6. Mix K_3PO_4 and KH_2PO_4 in the proper proportion.

8.44   The pH of blood plasma is 7.40. Assuming the principal buffer system is HCO_3^-/H_2CO_3, calculate the ratio $[HCO_3^-]/[H_2CO_3]$. Is this buffer more effective against an added acid or an added base?

$$pH = pK_a + \log \frac{\left[HCO_3^-\right]}{\left[H_2CO_3\right]}$$

$$7.40 = 6.38 + \log \frac{\left[HCO_3^-\right]}{\left[H_2CO_3\right]}$$

$$\frac{\left[HCO_3^-\right]}{\left[H_2CO_3\right]} = 10^{1.02} = 10.5$$

The buffer should be more effective against an added acid because approximately ten times more base is present compared to acid.

8.46   The buffer range is defined by the equation $pH = pK_a \pm 1$. Calculate the range of the ratio [conjugate base]/[acid] that corresponds to this equation.

When $pK_a = pH - 1$,

$$pH = pK_a + \log \frac{\left[\text{conjugate base}\right]}{\left[\text{acid}\right]}$$

$$\log \frac{\left[\text{conjugate base}\right]}{\left[\text{acid}\right]} = 1$$

$$\frac{\left[\text{conjugate base}\right]}{\left[\text{acid}\right]} = 10$$

When $pK_a = pH + 1$,

$$pH = pK_a + \log \frac{[\text{conjugate base}]}{[\text{acid}]}$$

$$\log \frac{[\text{conjugate base}]}{[\text{acid}]} = -1$$

$$\frac{[\text{conjugate base}]}{[\text{acid}]} = 0.1$$

Therefore, the range of the ratio is $0.1 < \dfrac{[\text{conjugate base}]}{[\text{acid}]} < 10$.

8.48   How many milliliters of $1.0\ M$ NaOH must be added to 200 mL of $0.10\ M$ NaH_2PO_4 to make a buffer solution with a pH of 7.50?

Let x L of NaOH be the volume required. The number of moles of NaOH and NaH_2PO_4 before the reaction are

$$\text{Number of moles of NaOH} = (1.0\ M)\ (x\ \text{L}) = 1.0x\ \text{mol}$$

$$\text{Number of moles of } NaH_2PO_4 = (0.10\ M)\ (0.200\ \text{L}) = 0.0200\ \text{mol}$$

NaOH reacts essentially completely with $H_2PO_4^-$ according to the reaction

$$OH^- + H_2PO_4^- \rightarrow H_2O + HPO_4^{2-}$$

After the reaction,

$$\text{Number of moles of } H_2PO_4^- = (0.0200 - 1.0x)\ \text{mol}$$

$$\text{Number of moles of } HPO_4^{2-} = 1.0x\ \text{mol}$$

The solution now is a buffer solution. The pH of this solution is related to the concentrations of $H_2PO_4^-$ and HPO_4^{2-}:

$$pH = pK_a + \log \frac{\left[HPO_4^{2-}\right]}{\left[H_2PO_4^-\right]}$$

Since the acid/base pair is in a single solution, the concentration ratio is the same as the mole ratio:

$$\text{pH} = \text{p}K_a + \log \frac{\text{Number of moles of HPO}_4^{2-}}{\text{Number of moles of H}_2\text{PO}_4^-}$$

$$7.50 = 7.21 + \log \frac{1.0x \text{ mol}}{(0.0200 - 1.0x) \text{ mol}}$$

$$\frac{1.0x}{0.0200 - 1.0x} = 10^{0.29} = 1.95$$

$$1.0x = 1.95(0.0200 - 1.0x) = 0.0390 - 1.95x$$

$$x = 1.3 \times 10^{-2}$$

Therefore, a volume of 1.3×10^{-2} L, or 13 mL is required to prepare the buffer.

8.50 How would you prepare a CH$_3$COONa/CH$_3$COOH buffer with a pH of 4.40 and an ionic strength of 0.050 m? Treat molarity the same as molality.

The ionic strength is a result of the dissociation of CH$_3$COONa into Na$^+$ and CH$_3$COO$^-$. CH$_3$COOH is too weak an acid to contribute to the ionic strength. The molality of Na$^+$, m_+, is the same as that of CH$_3$COO$^-$, m_-.

$$I = \frac{1}{2}\left(m_+ z_+^2 + m_- z_-^2\right)$$

$$0.050 \, m = \frac{1}{2}\left[m_+ (1)^2 + m_- (1)^2\right] = m_-$$

Therefore, assuming molarity is the same as molarity, [CH$_3$COO$^-$] = 0.050 M in the buffer. Using this concentration and the Henderson–Hasselbalch equation, [CH$_3$COOH] can be calculated.

$$\text{pH} = \text{p}K_a + \log \frac{\left[\text{CH}_3\text{COO}^-\right]}{\left[\text{CH}_3\text{COOH}\right]}$$

$$4.40 = 4.76 + \log \frac{0.050}{\left[\text{CH}_3\text{COOH}\right]}$$

$$\frac{0.050}{\left[\text{CH}_3\text{COOH}\right]} = 10^{-0.36} = 0.437$$

$$\left[\text{CH}_3\text{COOH}\right] = 0.11 \, M$$

The buffer should contain 0.11 M CH$_3$COOH and 0.050 M CH$_3$COONa.

8.52 Which of the amino acids listed in Table 8.4 have a buffer capacity in the physiological region of pH 7?

An amino acid must have a pK_a (pK_a', pK_a'', or pK_a''') in the range of $6-8$ to exhibit buffer capacity in the physiological region of pH 7. Histidine, with p$K_a''' = 6.00$, is the only such amino acid.

8.54 From the pK_a values listed in Table 8.4, calculate the pI value for amino acids lysine and valine.

Lysine

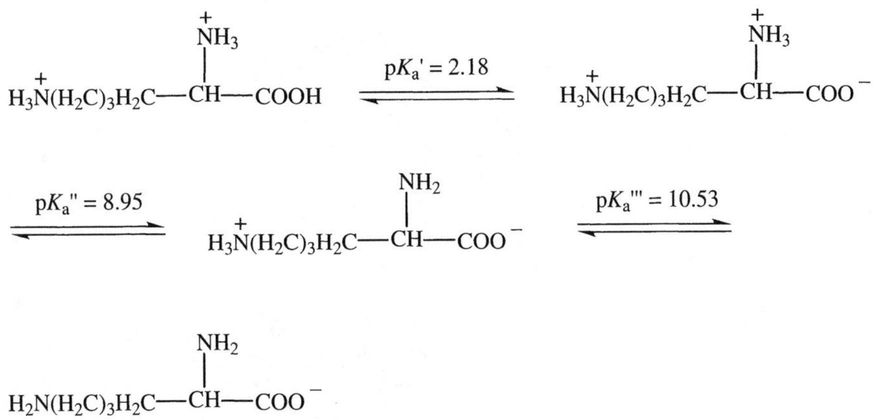

$$pI = \frac{8.95 + 10.53}{2} = 9.74$$

Valine

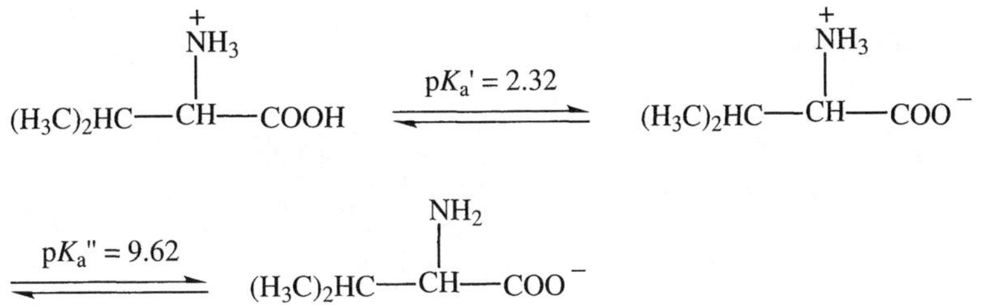

$$pI = \frac{2.32 + 9.62}{2} = 5.97$$

8.56 At neutral pH, amino acids exist as dipolar ions. Using glycine as an example, and given that the pK_a of the carboxyl group is 2.3 and that of the ammonium group is 9.6, predict the predominant form of the molecule at pHs of 1, 7, and 12. Justify your answers using Equation 8.16.

Use the Henderson–Hasselbalch equation,

$$pH = pK_a + \log \frac{[\text{conjugate base}]}{[\text{acid}]},$$

to calculate the ratios of each acid-conjugate base pair in glycine, NH_2-CH_2-COOH.

At pH = 1,

For the –COOH group,

$$1 = 2.3 + \log \frac{[-COO^-]}{[-COOH]}$$

$$-1.3 = \log \frac{[-COO^-]}{[-COOH]}$$

$$\frac{[-COO^-]}{[-COOH]} = 5.0 \times 10^{-2}$$

For the –NH$_2$ group,

$$1 = 9.6 + \log \frac{[-NH_2]}{[NH_3^+]}$$

$$-8.6 = \log \frac{[-NH_2]}{[-NH_3^+]}$$

$$\frac{[-NH_2]}{[-NH_3^+]} = 2.5 \times 10^{-9}$$

Thus, the predominant species is $NH_3^+-CH_2-COOH$.

At pH = 7,

For the –COOH group,

$$7 = 2.3 + \log \frac{[-COO^-]}{[-COOH]}$$

$$4.7 = \log \frac{[-COO^-]}{[-COOH]}$$

$$\frac{[-COO^-]}{[-COOH]} = 5.0 \times 10^{4}$$

For the –NH$_2$ group,

$$7 = 9.6 + \log \frac{[-NH_2]}{[NH_3^+]}$$

$$-2.6 = \log \frac{[-NH_2]}{[-NH_3^+]}$$

$$\frac{[-NH_2]}{[-NH_3^+]} = 2.5 \times 10^{-3}$$

Thus, the predominant species is $NH_3^+-CH_2-COO^-$.

At pH = 12,

For the –COOH group,

$$12 = 2.3 + \log \frac{[-COO^-]}{[-COOH]}$$

$$9.7 = \log \frac{[-COO^-]}{[-COOH]}$$

$$\frac{[-COO^-]}{[-COOH]} = 5.0 \times 10^9$$

For the –NH$_2$ group,

$$12 = 9.6 + \log \frac{[-NH_2]}{[-NH_3^+]}$$

$$2.4 = \log \frac{[-NH_2]}{[-NH_3^+]}$$

$$\frac{[-NH_2]}{[-NH_3^+]} = 2.5 \times 10^2$$

Thus, the predominant species is NH$_2$–CH$_2$–COO$^-$.

8.58 From the dependence of K_w on temperature (see p. 269), calculate the enthalpy of dissociation for water.

There are two ways to solve this problem. One uses the van't Hoff equation, and the other uses a graphical method.

Method 1

Substitute $K_w = 0.12 \times 10^{-14}$ at 273 K and $K_w = 5.4 \times 10^{-13}$ at 373 K into the van't Hoff equation:

$$\ln \frac{K_2}{K_1} = \frac{\Delta_r H^\circ}{R} \left(\frac{1}{T_1} - \frac{1}{T_2} \right)$$

$$\ln \frac{5.4 \times 10^{-13}}{0.12 \times 10^{-14}} = \frac{\Delta_r H^\circ}{8.314 \, \text{J K}^{-1} \, \text{mol}^{-1}} \left(\frac{1}{273 \, \text{K}} - \frac{1}{373 \, \text{K}} \right)$$

$$\Delta_r H^\circ = 5.2 \times 10^4 \, \text{J mol}^{-1}$$

Method 2

The slope of a plot of $\ln K_w$ vs $1/T$ has a value of $-\Delta_r H^\circ / R$. The following data points are used.

T/K	K_w	$10^3\,K/T$	$\ln K_w$
273	0.12×10^{-14}	3.663	-34.36
298	1.0×10^{-14}	3.356	-32.24
313	2.9×10^{-14}	3.195	-31.17
373	5.4×10^{-13}	2.681	-28.25

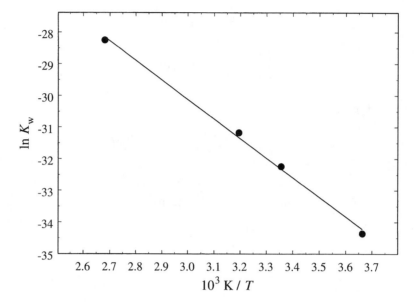

The slope of the line is -6.2×10^3 K.

$$\frac{-\Delta_r H^\circ}{R} = -6.2 \times 10^3 \text{ K}$$

$$\Delta_r H^\circ = \left(-6.2 \times 10^3 \text{ K}\right) \left(8.314 \text{ J K}^{-1}\,\text{mol}^{-1}\right)$$

$$= 5.2 \times 10^4 \text{ J mol}^{-1}$$

The two methods give the same answer, but the second method is preferred since it uses all the data and improves the statistical significance of the result. It is less likely to give a result that is skewed by a single bad data point as would happen if that point were chosen as one of the two points used in the first method.

8.60　Show that the acid dissociation constant, K_a, of a weak monoprotic acid in water is related to its concentration, c (mol L^{-1}), and its degree of dissociation, α, by $K_a = \alpha^2 c/(1-\alpha)$ if the self-dissociation of water is ignored. If the latter is taken into account, show that $K_a = \frac{1}{2}\alpha^2 c \left[1 + \left(1 + 4K_w\alpha^{-2}c^{-2}\right)^{1/2}\right] / (1-\alpha)$.

If the self dissociation of water is ignored, only the following equation needs to be considered:

	HA	$\rightleftharpoons$	H$^+$	$+$	A$^-$	
Initial	c		0		0	M
At equilibrium	$c(1-\alpha)$		$c\alpha$		$c\alpha$	M

The equilibrium expression is

$$K_a = \frac{(c\alpha)\,(c\alpha)}{c\,(1-\alpha)} = \frac{\alpha^2 c}{1-\alpha}$$

If the self dissociation of water is taken into account, the concentrations of HA and A⁻ are still the same as above at equilibrium, that is,

$$[HA] = c\,(1 - \alpha)$$

$$\left[A^-\right] = c\alpha$$

However, [H⁺] is derived both from HA and H_2O. It can be calculated by considering charge balance:

$$\left[H^+\right] = \left[OH^-\right] + \left[A^-\right]$$

An expression relating [H⁺], K_w, c, and α is obtained by solving for [OH⁻] and then substituting into the equilibrium expression for the self dissociation of water.

$$K_w = \left[H^+\right]\left[OH^-\right] = \left[H^+\right]\left(\left[H^+\right] - \left[A^-\right]\right) = \left[H^+\right]\left(\left[H^+\right] - c\alpha\right)$$

$$\left[H^+\right]^2 - c\alpha\left[H^+\right] - K_w = 0$$

$$\left[H^+\right] = \frac{c\alpha \pm \sqrt{c^2\alpha^2 + 4K_w}}{2}$$

Since $\sqrt{c^2\alpha^2 + 4K_w} > c\alpha$, the only physically possible root is

$$\left[H^+\right] = \frac{c\alpha + \sqrt{c^2\alpha^2 + 4K_w}}{2}$$

The equilibrium expression for the acid dissociation is

$$K_a = \frac{\left[H^+\right]\left[A^-\right]}{[HA]}$$

$$= \frac{\left(\frac{c\alpha + \sqrt{c^2\alpha^2 + 4K_w}}{2}\right)(c\alpha)}{c\,(1 - \alpha)}$$

$$= \frac{\alpha}{2\,(1 - \alpha)}\left[c\alpha + \sqrt{c^2\alpha^2\left(1 + 4K_w c^{-2}\alpha^{-2}\right)}\right]$$

$$= \frac{\alpha}{2\,(1 - \alpha)}(c\alpha)\left[1 + \left(1 + 4K_w c^{-2}\alpha^{-2}\right)^{1/2}\right]$$

$$= \frac{\alpha^2 c}{2\,(1 - \alpha)}\left[1 + \left(1 + 4K_w c^{-2}\alpha^{-2}\right)^{1/2}\right]$$

8.62 Depending on the pH of the solution, ferric ions (Fe^{3+}) may exist in the free-ion form or form the insoluble precipitate $Fe(OH)_3$ ($K_{sp} = 1.0 \times 10^{-36}$). Calculate the pH at which 90% of the Fe^{3+} ions in a $4.5 \times 10^{-5}\,M$ Fe^{3+} solution would be precipitated. What conclusion can you draw about the Fe^{3+} ion concentration in blood plasma whose pH is 7.40?

The concentration of Fe^{3+} remaining in the solution is

$$\left[Fe^{3+}\right] = (10\%)\left(4.5 \times 10^{-5}\,M\right) = 4.5 \times 10^{-6}\,M$$

The solubility equilibrium of $Fe(OH)_3$ is

$$Fe(OH)_3(s) \rightleftharpoons Fe^{3+}(aq) + 3OH^-(aq)$$

and $[OH^-]$ can be calculated from the equilibrium expression

$$K_{sp} = 1.0 \times 10^{-36} = \left[Fe^{3+}\right]\left[OH^-\right]^3 = \left(4.5 \times 10^{-6}\right)\left[OH^-\right]^3$$

$$\left[OH^-\right] = 6.06 \times 10^{-11}\ M$$

Therefore,

$$\left[H^+\right] = \frac{K_w}{\left[OH^-\right]} = \frac{1.00 \times 10^{-14}}{6.06 \times 10^{-11}} = 1.65 \times 10^{-4}\ M$$

$$pH = 3.783$$

In blood plasma, where pH = 7.40, and $[H^+] = 4.0 \times 10^{-8}\ M$,

$$[OH^-] = \frac{K_w}{[H^+]} = \frac{2.1 \times 10^{-14}}{4.0 \times 10^{-8}} = 5.3 \times 10^{-7}\ M$$

where the value of K_w appropriate for the physiological temperature of 37° C is used. Thus, the free Fe^{3+} ion concentration in blood plasma is limited by the solubility of $Fe(OH)_3$, whose K_{sp} is assumed to be temperature independent.

$$[Fe^{3+}] \leq \frac{K_{sp}}{[OH^-]^3}$$

$$= \frac{1.0 \times 10^{-36}}{\left(5.3 \times 10^{-7}\right)^3}$$

$$= 6.7 \times 10^{-18}\ M$$

There is virtually no free Fe^{3+} in blood plasma.

8.64   The pH of gastric juice is about 1.00 and blood plasma is 7.40. Calculate the Gibbs energy required to secrete a mole of H^+ ions from blood plasma to the stomach at 37° C. Assume ideal behavior.

In the blood plasma, $[H^+] = 10^{-pH} = 10^{-7.40} = 4.0 \times 10^{-8}\ M$, and in the gastric juice, $[H^+] = 10^{-pH} = 10^{-1.00} = 1.0 \times 10^{-1}\ M$. The "reaction" is

$$H^+(4.0 \times 10^{-8}\ M) \rightarrow H^+(1.0 \times 10^{-1}\ M)$$

Since $\Delta G = \Delta G° + RT \ln Q$ and $\Delta G° = 0$ for this reaction,

$$\Delta G = RT \ln \frac{[\text{H}^+]_{\text{stomach}}}{[\text{H}^+]_{\text{blood plasma}}}$$

$$= \left(8.314 \text{ J K}^{-1} \text{ mol}^{-1}\right)(310 \text{ K}) \ln \frac{1.0 \times 10^{-1}}{4.0 \times 10^{-8}}$$

$$= 3.8 \times 10^4 \text{ J mol}^{-1}$$

$$= 38 \text{ kJ mol}^{-1}$$

8.66   Calcium oxalate is a major component of kidney stones. From the dissociation constants listed in Table 8.1 and given that the solubility product of CaC_2O_4 is 3.0×10^{-9}, predict whether the formation of kidney stones can be minimized by increasing or decreasing the pH of the fluid present in the kidneys. The pH of normal kidney fluid is about 8.2.

The solubility equilibrium of CaC_2O_4 is

$$CaC_2O_4(s) \rightleftharpoons Ca^{2+}(aq) + C_2O_4^{2-}(aq)$$

The resulting oxalate ions hydrolyze:

$$C_2O_4^{2-}(aq) + 2H_2O(l) \rightleftharpoons H_2C_2O_4(aq) + 2OH^-(aq)$$

Increasing the pH of the fluid present in the kidneys will increase $[\text{OH}^-]$, which, according to Le Chatelier's principle, will shift the equilibrium towards $C_2O_4^{2-}$, which will in turn shift the solubility equilibrium of calcium oxalate towards CaC_2O_4. In other words, increasing the pH will enhance the formation of kidney stones. On the other hand, decreasing the pH of the fluid will decrease $[\text{OH}^-]$, which will cause the hydrolysis equilibrium to shift to the right. The removal of $C_2O_4^{2-}$ as $H_2C_2O_4$ will cause the solubility equilibrium of calcium oxalate to shift to the right. As a result, decreasing the pH of the fluid will minimize kidney stone formation.

8.68   From the dissociation constant of formic acid listed in Table 8.1, calculate the Gibbs energy and the standard Gibbs energy for the dissociation of formic acid at 298 K.

The reaction is

$$\text{HCOOH} \rightleftharpoons \text{H}^+ + \text{HCOO}^-$$

At equilibrium, the Gibbs energy, $\Delta_r G = 0$. The standard Gibbs energy is related to the dissociation constant.

$$\Delta_r G^\circ = -RT \ln K_a$$

$$= -\left(8.314 \text{ J K}^{-1} \text{ mol}^{-1}\right)(298 \text{ K}) \ln\left(1.77 \times 10^{-4}\right) = 2.14 \times 10^4 \text{ J mol}^{-1}$$

8.70   A 0.400 M formic acid (HCOOH) solution freezes at $-0.758°$ C. Calculate the value of K_a at that temperature. (*Hint*: Assume that molarity is equal to molality.)

The dissociation reaction is

$$\begin{array}{ccccc} & HCOOH(aq) & \rightleftharpoons & H^+(aq) & + & HCOO^-(aq) \\ \text{Initial} & 0.400 & & 0 & & 0 & M \\ \text{At equilibrium} & 0.400 - x & & x & & x & M \end{array}$$

Assuming molarities are equal to molalities, the total molality of all species is

$$(0.400 - x + x + x)\ m = (0.400 + x)\ m$$

This molality is related to the freezing point of the solution.

$$\Delta T_f = K_f m$$

$$0.758\ K = \left(1.86\ K\,mol^{-1}\,kg\right)(0.400 + x)\ mol\,kg^{-1}$$

$$x = 0.0075\ m = 0.0075\ M$$

Knowing the value of x, the dissociation constant can be calculated.

$$K_a = \frac{x^2}{0.400 - x} = 1 \times 10^{-4}$$

The result is limited by the precision to which the concentration of the formic acid solution is known at equilibrium.

8.72 Acid–base reactions usually go to completion. Confirm this statement by calculating the equilibrium constant for each of the following cases: **(a)** a strong acid reacting with a strong base, **(b)** a strong acid reacting with a weak base (NH_3), **(c)** a weak acid (CH_3COOH) reacting with a strong base, and **(d)** a weak acid (CH_3COOH) reacting with a weak base (NH_3). (*Hint*: Strong acids exist as H^+ ions, and strong bases exist as OH^- ions in solution. You need to look up K_a, K_b, and K_w values.)

(a) The reaction is $H^+ + OH^- \rightleftharpoons H_2O$ for which $K = \dfrac{1}{K_w} = 1.0 \times 10^{14}$.

(b) The reaction is $H^+ + NH_3 \rightleftharpoons NH_4^+$ for which $K = \dfrac{1}{K_a} = \dfrac{1}{5.6 \times 10^{-10}} = 1.8 \times 10^9$.

(c) The reaction,

$$CH_3COOH + OH^- \rightleftharpoons CH_3COO^- + H_2O$$

may be considered as the sum of the two equations, each with its own equilibrium constant

$$CH_3COOH \rightleftharpoons CH_3COO^- + H^+ \quad K_a$$

$$H^+ + OH^- \rightleftharpoons H_2O \quad 1/K_w$$

The overall equilibrium constant is

$$K = K_a \left(1/K_w\right) = \frac{K_a}{K_w} = \frac{1.8 \times 10^{-5}}{1.0 \times 10^{-14}} = 1.8 \times 10^9$$

(d) Again, the reaction,

$$CH_3COOH + NH_3 \rightleftharpoons CH_3COO^- + NH_4^+$$

may be considered as the sum of the two equations, each with its own equilibrium constant

$$CH_3COOH \rightleftharpoons CH_3COO^- + H^+ \quad K_a$$

$$NH_3 + H^+ \rightleftharpoons NH_4^+ \quad 1/K_a'$$

The overall equilibrium constant is

$$K = K_a\left(1/K_a'\right) = \frac{K_a}{K_a'} = \frac{1.8 \times 10^{-5}}{5.6 \times 10^{-10}} = 3.2 \times 10^4$$

In each case the overall equilibrium constant is large enough that the reaction proceeds to completion.

8.74   One of the most common antibiotics is penicillin G (benzylpenicillinic acid), which has the following structure:

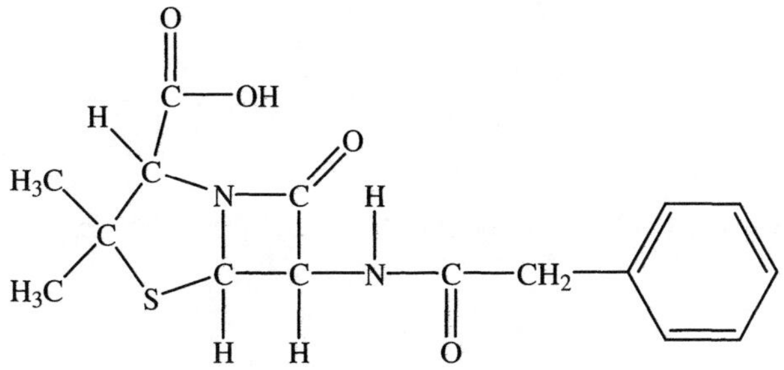

It is a weak monoprotic acid:

$$HP \rightleftharpoons H^+ + P^- \qquad K_a = 1.64 \times 10^{-3}$$

where HP denotes the parent acid and P^- the conjugate base. Penicillin G is produced by growing molds in fermentation tanks at 25° C and a pH range of 4.5 to 5.0. The crude form of this antibiotic is obtained by extracting the fermentation broth with an organic solvent in which the acid is soluble. **(a)** Identify the acidic hydrogen atom. **(b)** In one stage of purification, the organic extract of the crude penicillin G is treated with a buffer solution at pH = 6.50. What is the ratio of the conjugate base of penicillin G to the acid at this pH? Would you expect the conjugate base to be more soluble in water than the acid? **(c)** Penicillin G is not suitable for oral administration, but the sodium salt (NaP) is because it is soluble. Calculate the pH of a 0.12 M NaP solution formed when a tablet containing the salt is dissolved in a glass of water.

(a) The acidic hydrogen is from the carboxyl group (–COOH).

(b) The ratio of the conjugate base and the acid can be obtained from the Henderson–Hasselbalch equation:

$$pH = pK_a + \log \frac{[P^-]}{[HP]}$$

$$6.50 = -\log\left(1.64 \times 10^{-3}\right) + \log \frac{[P^-]}{[HP]}$$

$$\frac{[P^-]}{[HP]} = 10^{3.715} = 5.2 \times 10^3$$

Thus, nearly all of the penicillin G will be in the ionized form. The ionized form is more soluble in water because it bears a net charge; penicillin G is largely nonpolar and therefore much less soluble in water. (Both penicillin G and its salt are effective antibiotics.)

(c) NaP dissociates into Na^+ and P^-, and P^- undergoes hydrolysis:

	$P^-(aq)$	$+$	$H_2O(l)$	$\rightleftharpoons$	$HP(aq)$	$+$	OH^-	
Initial	0.12				0		0	M
At equilibrium	$0.12 - x$				x		x	M

The equilibrium expression associated with the hydrolysis reaction is

$$K_b = \frac{K_w}{K_a} = \frac{x^2}{0.12 - x}$$

$$\frac{1.00 \times 10^{-14}}{1.64 \times 10^{-3}} = 6.098 \times 10^{-12} = \frac{x^2}{0.12 - x}$$

Since K_b is very small, $0.12 - x$ can be approximated by 0.12. The expression above simplifies to

$$6.098 \times 10^{-12} = \frac{x^2}{0.12}$$

$$x = 8.55 \times 10^{-7}$$

Checking the assumption,

$$\frac{x}{0.12} \times 100\% = 7.1 \times 10^{-4}\% < 5\%$$

Thus, the assumption is valid.

Therefore,

$$[H^+] = \frac{K_w}{[OH^-]} = \frac{1.00 \times 10^{-14}}{8.55 \times 10^{-7}} = 1.17 \times 10^{-8}$$

$$pH = 7.93$$

Because HP is a relatively strong acid, P^- is a weak base. Consequently, only a small fraction of P^- undergoes hydrolysis and the solution is slightly basic.

8.76 Referring to the buffer system listed in Problem 8.36, calculate the pH of the buffer after it has been diluted by a factor of 100. What would be the change in pH if the base component of the buffer were diluted by the same factor?

Dilution of a buffer

After the buffer has been diluted by a factor of 100, $[NH_3] = 0.0020\ M$ and $[NH_3] = 0.0020\ M$. The pH of the buffer can be calculated using the Henderson–Hasselbalch equation.

$$pH = pK_a + \log\left(\frac{[NH_3]}{[NH_4^+]}\right)$$

$$= 9.25 + \log\left(\frac{0.0020}{0.0020}\right) = 9.25$$

The dilution of a buffer does not change its pH.

Dilution of a basic solution

First calculate the pH of the 0.20-M solution.

	NH_3	+	H_2O	$\rightleftharpoons$	NH_4^+	+	OH^-	
Initial	0.20				0		0	M
At equilibrium	0.20 − x				x		x	M

$$K_b = 1.8 \times 10^{-5} = \frac{[NH_4^+][OH^-]}{[NH_3]} = \frac{x^2}{0.20 - x}$$

Since K_b is small, $0.20 - x$ can be approximated as 0.20, and the expression above simplies to

$$1.8 \times 10^{-5} = \frac{x^2}{0.20}$$

$$x = 1.90 \times 10^{-3}$$

Checking the assumption,

$$\frac{x}{0.20} \times 100\% = 0.95\%$$

Thus, the assumption is valid. The pH of the solution is

$$pH = 14.00 - pOH = 14.00 - \left[-\log\left(1.90 \times 10^{-3}\right)\right] = 11.28$$

Now dilute the basic solution by a factor of 100.

	NH_3	+	H_2O	$\rightleftharpoons$	NH_4^+	+	OH^-	
Initial	0.0020				0		0	M
At equilibrium	0.0020 − x				x		x	M

$$K_b = 1.8 \times 10^{-5} = \frac{\left[NH_4^+\right]\left[OH^-\right]}{\left[NH_3\right]} = \frac{x^2}{0.0020 - x}$$

$$x^2 + 1.8 \times 10^{-5}x - 3.6 \times 10^{-8} = 0$$

$$x = 1.81 \times 10^{-4}$$

The nonphysical root, with a minus sign preceding the square root, has been discarded. The pH of the solution is

$$pH = 14.00 - pOH = 14.00 - \left[-\log\left(1.81 \times 10^{-4}\right)\right] = 10.26$$

There is a decrease of 1.02 pH unit when the basic solution is diluted by a factor of 100.

8.78   Tris [tris(hydroxymethyl)aminomethane] is a widely used buffer by biochemists. It has a pK_a of 8.30 at 20° C. Calculate the buffer capacity of a 0.10 M Tris buffer (containing both Tris and its conjugate acid TrisH$^+$) at **(a)** pH = 8.30 and **(b)** pH = 10.30. In each case 0.020 mol H$^+$ ions is added to one liter of the buffer solution.

The pH of the buffer is given by the Henderson–Hasselbalch equation:

$$pH = pK_a + \log \frac{[Tris]}{\left[TrisH^+\right]} = 8.30 + \log \frac{[Tris]}{\left[TrisH^+\right]}$$

Since

$$[Tris] + \left[TrisH^+\right] = 0.10 \ M$$

The Henderson–Hasselbalch equation can be rewritten as

$$pH = 8.30 + \log \frac{0.10 - \left[TrisH^+\right]}{\left[TrisH^+\right]}$$

Once the pH of the buffer is known, the concentrations of Tris and TrisH$^+$ can be determined.

(a) For a pH = 8.30 buffer,

$$pH = 8.30 = 8.30 + \log \frac{0.10 - \left[TrisH^+\right]}{\left[TrisH^+\right]}$$

$$\log \frac{0.10 - \left[TrisH^+\right]}{\left[TrisH^+\right]} = 0$$

$$\frac{0.10 - \left[TrisH^+\right]}{\left[TrisH^+\right]} = 1$$

$$\left[TrisH^+\right] = 0.050 \ M$$

$$[Tris] = 0.10 - \left[TrisH^+\right] = 0.050 \ M$$

Buffer capacity is defined in terms of 1-L of buffer. The concentration of H$^+$ added to the buffer is therefore $\frac{0.020 \text{ mol}}{1 \text{ L}} = 0.020 \ M$.

The reaction of H^+ with Tris can be represented by

$$H^+ + Tris \rightleftharpoons TrisH^+$$

and will proceed to near completion. Therefore, after the reaction,

$$[Tris] = 0.050\ M - 0.020\ M = 0.030\ M$$

$$\left[TrisH^+\right] = 0.050\ M + 0.020\ M = 0.070\ M$$

The change in the concentration of the base component of the buffer, Tris, is $-0.020\ M$. The pH of the Tris buffer after the addition of H^+ can be calculated:

$$pH = pK_a + \log \frac{[Tris]}{\left[TrisH^+\right]}$$

$$= 8.30 + \log \frac{0.030}{0.070}$$

$$= 7.932$$

The buffer capacity is

$$\beta = \frac{-0.020\ M}{(7.932 - 8.30)\ pH\ unit} = 0.054\ M\ (pH\ unit)^{-1}$$

(b) For a pH = 10.30 buffer,

$$pH = 10.30 = 8.30 + \log \frac{0.10 - \left[TrisH^+\right]}{\left[TrisH^+\right]}$$

$$\log \frac{0.10 - \left[TrisH^+\right]}{\left[TrisH^+\right]} = 2$$

$$\frac{0.10 - \left[TrisH^+\right]}{\left[TrisH^+\right]} = 100$$

$$\left[TrisH^+\right] = 9.90 \times 10^{-4}\ M$$

$$[Tris] = 0.10 - \left[TrisH^+\right] = 0.0990\ M$$

The reaction of H^+ with Tris will proceed to near completion. Therefore, in 1-L of buffer, after the reaction,

$$[Tris] = 0.0990\ M - 0.020\ M = 0.0790\ M$$

$$\left[TrisH^+\right] = 9.90 \times 10^{-4}\ M + 0.020\ M = 0.0210\ M$$

Once again, the change in the concentration of the base component of the buffer, Tris, is $-0.020\ M$. The pH of the Tris buffer after the addition of H^+ can be calculated:

$$pH = pK_a + \log \frac{[Tris]}{\left[TrisH^+\right]}$$

$$= 8.30 + \log \frac{0.0790}{0.0210}$$

$$= 8.875$$

The buffer capacity is

$$\beta = \frac{-0.020 \ M}{(8.875 - 10.30) \ \text{pH unit}} = 0.014 \ M \ (\text{pH unit})^{-1}$$

The buffer has a greater buffer capacity when its pH is approximately the same as the pK_a of its acidic component.

8.80 The pK_a of the Tris buffer is 8.30 at 20° C. What is its value at 37° C? The molar enthalpy of dissociation for Tris is 48.0 kJ mol^{-1}.

The van't Hoff equation is used to calculate pK_a at 37° C.

$$\ln \frac{K_2}{K_1} = \frac{\Delta_r H^\circ}{R} \left(\frac{1}{T_1} - \frac{1}{T_2} \right)$$

$$\log \frac{K_2}{K_1} = \frac{\Delta_r H^\circ}{2.303R} \left(\frac{1}{T_1} - \frac{1}{T_2} \right)$$

$$\left[-\log K_1 - (-\log K_2) \right] = pK_1 - pK_2 = \frac{\Delta_r H^\circ}{2.303R} \left(\frac{1}{T_1} - \frac{1}{T_2} \right)$$

$$8.30 - pK_2 = \frac{48.0 \times 10^3 \ \text{J mol}^{-1}}{2.303 \left(8.314 \ \text{J K}^{-1} \text{mol}^{-1} \right)} \left(\frac{1}{293.2 \ \text{K}} - \frac{1}{310.2 \ \text{K}} \right)$$

$$pK_2 = 7.83$$

TrisH$^+$ is a stronger acid at a higher temperature.

8.82 The volume of a 1 m glycine solution depends on the pH. At what pH would the volume be a minimum? Why? Assume temperature is kept constant.

In the zwitterionic form, the glycine molecule possesses two full charges, each of which can form ion-dipole bonds with water and drawing the solvent molecules closer. Consequently, at the isoelectric point (pH = pI = 5.97) where this form predominates, the ion-dipole forces will be strongest and the solution volume the smallest.

Chemical Kinetics

PROBLEMS AND SOLUTIONS

9.2 The rate law for the reaction

$$NH_4^+(aq) + NO_2^-(aq) \rightarrow N_2(g) + 2H_2O(l)$$

is given by rate = k [NH$_4^+$] [NO$_2^-$]. At 25° C, the rate constant is $3.0 \times 10^{-4}\ M^{-1}\,s^{-1}$. Calculate the rate of the reaction at this temperature if [NH$_4^+$] = 0.26 M and [NO$_2^-$] = 0.080 M.

$$\text{Rate} = k\,[NH_4^+]\,[NO_2^-] = \left(3.0 \times 10^{-4}\ M^{-1}\,s^{-1}\right)(0.26\ M)\,(0.080\ M) = 6.2 \times 10^{-6}\ M\,s^{-1}$$

9.4 The following reaction is found to be first order in A:

$$A \rightarrow B + C$$

If half of the starting quantity of A is used up after 56 s, calculate the fraction that will be used up after 6.0 min.

From the half-life (56 s), k can be determined:

$$k = \frac{\ln 2}{t_{1/2}} = \frac{\ln 2}{56\ s} = 1.24 \times 10^{-2}\ s^{-1}$$

The fraction of A that will remain after 6.0 min (3.6×10^2 s) is

$$\frac{[A]}{[A]_0} = e^{-kt} = e^{-\left(1.24 \times 10^{-2}\ s^{-1}\right)\left(3.6 \times 10^2\ s\right)} = 0.0115$$

Thus, the fraction that will be used up after 6.0 min is $1 - 0.0115 = 0.99$.

9.6 **(a)** The half-life of the first-order decay of radioactive ^{14}C is about 5720 years. Calculate the rate constant for the reaction. **(b)** The natural abundance of ^{14}C isotope is 1.1×10^{-13} mol % in living matter. Radiochemical analysis of an object obtained in an archaeological excavation shows that the ^{14}C isotope content is 0.89×10^{-14} mol %. Calculate the age of the object. State any assumptions.

(a)
$$k = \frac{\ln 2}{t_{1/2}} = \frac{\ln 2}{5720 \text{ yr}} = 1.212 \times 10^{-4} \text{ yr}^{-1} = 1.21 \times 10^{-4} \text{ yr}^{-1}$$

(b)
$$\frac{[^{14}C]}{[^{14}C]_0} = e^{-kt}$$

$$t = -\frac{1}{k} \ln \frac{[^{14}C]}{[^{14}C]_0}$$

Due to constant exchange of material with the surroundings, the mol % of ^{14}C of all living matter is assumed to be the same. However, when the object ceases to live, it no longer exchanges material with the environment and the mol % of ^{14}C will decrease according to first-order decay kinetics. Therefore, the ratio between $[^{14}C]$ and $[^{14}C]_0$ depends on the time elapsed since the object's "death." Thus, t in the equation above gives the age of the object.

$$t = -\frac{1}{1.212 \times 10^{-4} \text{ yr}^{-1}} \ln \frac{0.89 \times 10^{-14}}{1.1 \times 10^{-13}} = 2.1 \times 10^4 \text{ yr}$$

A key assumption in radiocarbon dating is that the natural abundance of ^{14}C has remained constant throughout the ages. Since the production of terrestrial ^{14}C is due to bombardment of ^{14}N by cosmic rays, variations in cosmic ray flux have in fact led to variations in the natural abundance of ^{14}C.

9.8   When the concentration of A in the reaction A → B was changed from 1.20 M to 0.60 M, the half-life increased from 2.0 min to 4.0 min at 25° C. Calculate the order of the reaction and the rate constant.

The half life is related to the initial concentration of A by

$$t_{1/2} \propto \frac{1}{[A]_0^{n-1}}$$

According to the data, the half-life doubled when $[A]_0$ was halved. This is only possible if the half-life is inversely proportional to $[A]_0$, or the reaction order, $n = 2$, indicating a second-order reaction.

The rate constant can be calculated using either $[A]_0 = 1.20$ M or 0.60 M and the corresponding half-life.

$$k = \frac{1}{[A]_0 t_{1/2}} = \frac{1}{(1.20 \text{ } M)(2.0 \text{ min})} = 0.42 \text{ } M^{-1} \text{ min}^{-1}$$

9.10   Cyclobutane decomposes to ethylene according to the equation

$$C_4H_8(g) \rightarrow 2C_2H_4(g)$$

Determine the order of the reaction and the rate constant based on the following pressures, which were recorded when the reaction was carried out at 430° C in a constant-volume vessel:

Time/s	$P_{C_4H_8}$/mmHg
0	400
2000	316
4000	248
6000	196
8000	155
10000	122

At constant temperature and volume, the pressure of cyclobutane is proportional to its concentration. A plot of $\ln P$ vs t shows a straight line with an equation of $y = -1.19 \times 10^{-4}x + 5.99$. Thus, the reaction is first order.

The slope of the line is $-k$. In other words, $k = 1.19 \times 10^{-4}\ \text{s}^{-1}$.

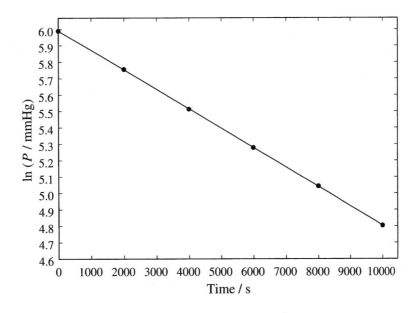

9.12 The rate constant for the second-order reaction

$$2NO_2(g) \rightarrow 2NO(g) + O_2(g)$$

is $0.54\ M^{-1}\,\text{s}^{-1}$ at 300° C. How long (in seconds) would it take for the concentration of NO_2 to decrease from $0.62\ M$ to $0.28\ M$?

$$\frac{1}{[A]} - \frac{1}{[A]_0} = kt$$

$$t = \frac{1}{k}\left(\frac{1}{[A]} - \frac{1}{[A]_0}\right) = \frac{1}{0.54\ M^{-1}\,\text{s}^{-1}}\left(\frac{1}{0.28\ M} - \frac{1}{0.62\ M}\right) = 3.6\ \text{s}$$

9.14   The integrated rate law for the zero-order reaction A → B is $[A] = [A]_0 - kt$. **(a)** Sketch the following plots: **(i)** rate versus [A] and **(ii)** [A] versus t. **(b)** Derive an expression for the half-life of the reaction. **(c)** Calculate the time in half-lives when the integrated rate law is no longer valid (that is, when [A] = 0).

(a) (i) The rate law is

$$Rate = k$$

Therefore, the rate of reaction is independent of the concentration of A.

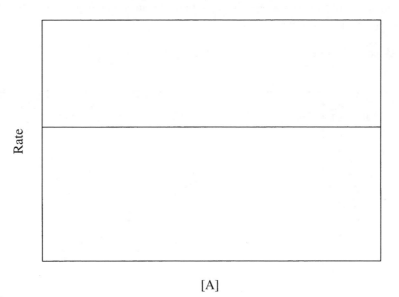

[A]

(a) (ii) Since

$$[A] = [A]_0 - kt$$

a plot of [A] vs t is a straight line with a slope of $-k$.

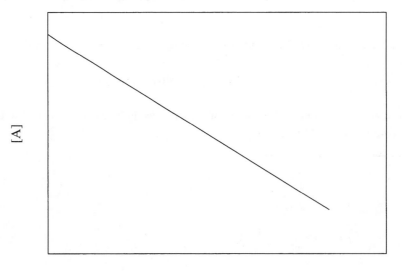

Time

(b) At $t = t_{1/2}$, $[A] = [A]_0/2$. Therefore,

$$\frac{[A]_0}{2} = [A]_0 - kt_{1/2}$$

$$t_{1/2} = \frac{1}{2k}[A]_0$$

(c) When $[A] = 0$,

$$[A] = 0 = [A]_0 - kt$$

According to part **(b)**,

$$k = \frac{1}{2t_{1/2}}[A]_0$$

Therefore, the time it takes to consume all the reactant is

$$t = \frac{[A]_0}{k} = \frac{[A]_0}{\frac{1}{2t_{1/2}}[A]_0} = 2t_{1/2}$$

The integrated rate law is no longer valid after 2 half-lives.

9.16 Many reactions involving heterogeneous catalysis are zero order; that is, rate $= k$. An example is the decomposition of phosphine (PH_3) over tungsten (W):

$$4PH_3(g) \rightarrow P_4(g) + 6H_2(g)$$

The rate for this reaction is independent of $[PH_3]$ as long as phosphine's pressure is sufficiently high (≥ 1 atm). Explain.

With sufficient PH_3, all the catalytic sites on the tungsten surface are occupied. Further increases in the amount of phosphine cannot affect the reaction, and the rate is independent of $[PH_3]$.

9.18 Consider the following nuclear decay

$$^{64}Cu \rightarrow {}^{64}Zn + {}^{0}_{-1}\beta \qquad t_{1/2} = 12.8 \text{ h}$$

Starting with one mole of ^{64}Cu, calculate the number of grams of ^{64}Zn formed after 25.6 hours.

Two half-lives would have elapsed after 25.6 hours. After the first half-life, 1/2 mole of ^{64}Cu would remain. After the second half-life, 1/4 mol of ^{64}Cu would remain. Thus, a total of 3/4 mole of ^{64}Cu decays to form 3/4 mole of ^{64}Zn. The mass of ^{64}Zn produced is

$$m = \left(\frac{3}{4} \text{ mol}\right)\left(63.93 \text{ g mol}^{-1}\right) = 47.95 \text{ g}$$

9.20 Derive Equation 9.27 using the steady-state approximation for both the H and Br atoms.

Apply the steady state approximation to both the H and Br atoms:

$$\frac{d\,[\text{H}]}{dt} = k_2[\text{Br}][\text{H}_2] - k_3[\text{H}][\text{Br}_2] - k_4[\text{H}][\text{HBr}] = 0 \tag{9.20.1}$$

$$\frac{d\,[\text{Br}]}{dt} = 2k_1[\text{Br}_2] - k_2[\text{Br}][\text{H}_2] + k_3[\text{H}][\text{Br}_2] + k_4[\text{H}][\text{HBr}] - k_5[\text{Br}]^2 = 0 \tag{9.20.2}$$

Summing Eqs. 9.20.1 and 9.20.2,

$$2k_1[\text{Br}_2] - k_5[\text{Br}]^2 = 0$$

$$[\text{Br}] = \left(\frac{2k_1[\text{Br}_2]}{k_5}\right)^{1/2} \tag{9.20.3}$$

Solving for [H] using Eq. 9.20.1,

$$[\text{H}] = \frac{k_2[\text{Br}][\text{H}_2]}{k_3[\text{Br}_2] + k_4[\text{HBr}]} \tag{9.20.4}$$

Substitute Eq. 9.20.3 into Eq. 9.20.4 to obtain an expression for [H] in terms of the concentrations of the reactants and products, and of the rate constants.

$$[\text{H}] = \frac{k_2 \left(\frac{2k_1}{k_5}\right)^{1/2} [\text{Br}_2]^{1/2}\,[\text{H}_2]}{k_3[\text{Br}_2] + k_4[\text{HBr}]} \tag{9.20.5}$$

According to the reaction mechanism,

$$\frac{d\,[\text{HBr}]}{dt} = k_2[\text{Br}][\text{H}_2] + k_3[\text{H}][\text{Br}_2] - k_4[\text{H}][\text{HBr}]$$

$$= k_2[\text{Br}][\text{H}_2] + [\text{H}]\left(k_3[\text{Br}_2] - k_4[\text{HBr}]\right) \tag{9.20.6}$$

Substitute Eqs. 9.20.3 and 9.20.5 into Eq. 9.20.6,

$$\frac{d\,[\text{HBr}]}{dt} = k_2\left(\frac{2k_1}{k_5}\right)^{1/2}[\text{Br}_2]^{1/2}\,[\text{H}_2] + \frac{k_2\left(\frac{2k_1}{k_5}\right)^{1/2}[\text{Br}_2]^{1/2}\,[\text{H}_2]}{k_3[\text{Br}_2] + k_4[\text{HBr}]}\left(k_3[\text{Br}_2] - k_4[\text{HBr}]\right)$$

$$= k_2\left(\frac{2k_1}{k_5}\right)^{1/2}[\text{Br}_2]^{1/2}\,[\text{H}_2]\left\{1 + \frac{k_3[\text{Br}_2] - k_4[\text{HBr}]}{k_3[\text{Br}_2] + k_4[\text{HBr}]}\right\}$$

$$= k_2\left(\frac{2k_1}{k_5}\right)^{1/2}[\text{Br}_2]^{1/2}\,[\text{H}_2]\left\{\frac{2k_3[\text{Br}_2]}{k_3[\text{Br}_2] + k_4[\text{HBr}]}\right\}$$

$$= k_2\left(\frac{2k_1}{k_5}\right)^{1/2}[\text{Br}_2]^{1/2}\,[\text{H}_2]\left\{\frac{2}{1 + \frac{k_4}{k_3}\frac{[\text{HBr}]}{[\text{Br}_2]}}\right\} \tag{9.20.7}$$

Setting

$$\alpha = 2k_2\left(\frac{2k_1}{k_5}\right)^{1/2}$$

$$\beta = \frac{k_4}{k_3}$$

Eq. 9.20.7 becomes

$$\frac{d\,[\text{HBr}]}{dt} = \frac{\alpha[\text{H}_2][\text{Br}_2]^{1/2}}{1 + \beta[\text{HBr}]/[\text{Br}_2]}$$

9.22 The following data were collected for the reaction between hydrogen and nitric oxide at 700° C:

$$2\text{H}_2(g) + 2\text{NO}(g) \rightarrow 2\text{H}_2\text{O}(g) + \text{N}_2(g)$$

Experiment	$[\text{H}_2]/M$	$[\text{NO}]/M$	Initial rate/$M \cdot \text{s}^{-1}$
1	0.010	0.025	2.4×10^{-6}
2	0.0050	0.025	1.2×10^{-6}
3	0.010	0.0125	0.60×10^{-6}

(a) What is the rate law for the reaction? **(b)** Calculate the rate constant for the reaction. **(c)** Suggest a plausible reaction mechanism that is consistent with the rate law. (*Hint:* Assume that the oxygen atom is the intermediate.) **(d)** More careful studies of the reaction show that the rate law over a wide range of concentrations of reactants should be

$$\text{rate} = \frac{k_1[\text{NO}]^2[\text{H}_2]}{1 + k_2[\text{H}_2]}$$

What happens to the rate law at very high and very low hydrogen concentrations?

(a) Comparing Experiment 1 and Experiment 2, the concentration of NO is constant and the concentration of H_2 has decreased by one-half. The initial rate has also decreased by one-half. Therefore, the initial rate is directly proportional to the concentration of H_2. That is, the reaction is first order in H_2.

Comparing Experiment 1 and Experiment 3, the concentration of H_2 is constant and the concentration of NO has decreased by one-half. The initial rate has decreased by one-fourth. Therefore, the initial rate is proportional to the squared concentration of NO. That is, the reaction is second order in NO.

Therefore, the rate law is

$$\text{Rate} = k[\text{NO}]^2[\text{H}_2]$$

(b) Using Experiment 1 to calculate the rate constant,

$$k = \frac{\text{Rate}}{[\text{NO}]^2[\text{H}_2]} = \frac{2.4 \times 10^{-6}\,M\,\text{s}^{-1}}{(0.025\,M)^2\,(0.010\,M)} = 0.38\,M^{-2}\,\text{s}^{-1}$$

(c) The rate law suggests that the slow step in the reaction mechanism will probably involve one H_2 molecule and two NO molecules. Additionally, the hint suggests that the O atom is an intermediate. A plausible mechanism is

$$\text{H}_2 + 2\text{NO} \longrightarrow \text{N}_2 + \text{H}_2\text{O} + \text{O} \quad \text{slow step}$$

$$\text{O} + \text{H}_2 \longrightarrow \text{H}_2\text{O} \quad\quad\quad\quad \text{fast step}$$

(d) At very high hydrogen concentrations, $k_2 [H_2] \gg 1$. Therefore, the rate law becomes

$$\text{rate} = \frac{k_1[NO]^2[H_2]}{k_2[H_2]} = \frac{k_1}{k_2}[NO]^2$$

At very low hydrogen concentrations, $k_2 [H_2] \ll 1$. Therefore, the rate law becomes

$$\text{rate} = k_1[NO]^2[H_2]$$

9.24   The gas-phase reaction between H_2 and I_2 to form HI involves a two-step mechanism:

$$I_2 \rightleftharpoons 2I$$

$$H_2 + 2I \rightarrow 2HI$$

The rate of formation of HI increases with the intensity of visible light. **(a)** Explain why this fact supports the two-step mechanism given. (*Hint*: The color of I_2 vapor is purple.) **(b)** Explain why the visible light has no effect on the formation of H atoms.

(a) In this two-step mechanism, the rate determining step is the second one where a hydrogen molecule collides with two iodine atoms. The absorption of visible light by the colored molecular iodine vapor weakens the I_2 bond and increases the number of I atoms present, which in turn increases the reaction rate.

(b) Hydrogen gas is colorless and does not absorb visible light. Ultraviolet light is required to photodissociate H_2 molecules.

9.26   Use Equation 9.23 to calculate the rate constant at 300 K for $E_a = 0$, 2, and 50 kJ mol^{-1}. Assume that $A = 10^{11}$ s^{-1} in each case.

For $E_a = 0$ kJ mol^{-1},

$$k = \left(10^{11} \text{ s}^{-1}\right) e^{-(0 \text{ J mol}^{-1})/(8.314 \text{J K}^{-1} \text{mol}^{-1})(300 \text{ K})} = 10^{11} \text{ s}^{-1}$$

For $E_a = 2$ kJ mol^{-1},

$$k = \left(10^{11} \text{ s}^{-1}\right) e^{-\left(2 \times 10^3 \text{ J mol}^{-1}\right)/(8.314 \text{J K}^{-1} \text{mol}^{-1})(300 \text{ K})} = 4.5 \times 10^{10} \text{ s}^{-1}$$

For $E_a = 50$ kJ mol^{-1},

$$k = \left(10^{11} \text{ s}^{-1}\right) e^{-\left(50 \times 10^3 \text{ J mol}^{-1}\right)/(8.314 \text{J K}^{-1} \text{mol}^{-1})(300 \text{ K})} = 2.0 \times 10^2 \text{ s}^{-1}$$

9.28   Over a range of about $\pm 3°$ C from normal body temperature the metabolic rate, M_T, is given by $M_T = M_{37} (1.1)^{\Delta T}$, where M_{37} is the normal rate and ΔT is the change in T. Discuss this equation in terms of a possible molecular interpretation. [*Source*: "Eco-Chem," J. A. Campbell, *J. Chem. Educ.* **52**, 327 (1975).]

Converting to kelvin, and using the Arrhenius equation,

$$\ln \frac{M_T}{M_{37}} = -\frac{E_a}{R} \left(\frac{1}{T} - \frac{1}{310 \text{ K}} \right)$$

Since the temperature range is so small, $f(T) = \frac{1}{T} - \frac{1}{310 \text{ K}}$ may be expanded in a Taylor series about $T_0 = 310$ K. Keeping only the first non-zero term results in $f(T) \approx -\frac{\Delta T}{T_0^2}$, where $\Delta T = 310 \text{ K} - T$. Thus,

$$\ln \frac{M_T}{M_{37}} = \frac{E_a}{R} \frac{\Delta T}{T_0^2}$$

or

$$M_T = M_{37} e^{\frac{E_a}{RT_0^2} \Delta T} = M_{37} \left(e^{\frac{E_a}{RT_0^2}} \right)^{\Delta T} = M_{37} (\text{constant})^{\Delta T}$$

which is of the observed form, providing an implicit factor of 1 K^{-1} is incorporated into the argument of the exponential function and the ΔT is interpreted as a unitless number. Specifically, it must be true that

$$e^{\frac{E_a}{RT_0^2}} = 1.1$$

$$\frac{E_a}{RT_0^2} = \ln 1.1 = 0.0953$$

$$E_a = \left(1 \text{ K}^{-1} \right) \left(8.314 \text{ J K}^{-1} \text{mol}^{-1} \right) (310 \text{ K})^2 (0.0953) = 7.6 \times 10^4 \text{ J mol}^{-1}$$

This activation energy is consistent with a single rate determining step controlling the metabolic rate within this temperature range.

9.30 The rate constants for the first-order decomposition of an organic compound in solution are measured at several temperatures:

k/s^{-1}	4.92×10^{-3}	0.0216	0.0950	0.326	1.15
$t/^\circ$ C	5.0	15	25	35	45

Determine graphically the pre-exponential factor and the energy of activation for the reaction.

Since

$$\ln k = \ln A - \frac{E_a}{RT}$$

A plot of $\ln k$ vs $1/T$ gives a slope of $-E_a/R$ and an intercept of $\ln A$. The following data are used for the plot:

10^3 K/T	3.595	3.470	3.353	3.245	3.143
$\ln(k/s^{-1})$	−5.314	−3.835	−2.354	−1.121	0.140

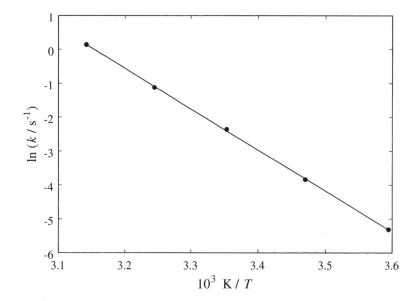

The equation for the line that best fits these points is $y = -1.207 \times 10^4 x + 38.06$. Therefore,

$$E_a = -\left(-1.207 \times 10^4 \text{ K}^{-1}\right)\left(8.314 \text{ J K}^{-1} \text{mol}^{-1}\right) = 1.00 \times 10^5 \text{ J mol}^{-1}$$

$$A = e^{38.06} = 3.38 \times 10^{16} \text{ s}^{-1}$$

9.32   The rate constant of a first-order reaction is 4.60×10^{-4} s^{-1} at 350° C. If the activation energy is 104 kJ mol^{-1}, calculate the temperature at which its rate constant is 8.80×10^{-4} s^{-1}.

$$\ln \frac{k_2}{k_1} = -\frac{E_a}{R}\left(\frac{1}{T_2} - \frac{1}{T_1}\right)$$

$$\ln \frac{8.80 \times 10^{-4}}{4.60 \times 10^{-4}} = -\frac{104 \times 10^3 \text{ J mol}^{-1}}{8.314 \text{ J K}^{-1} \text{mol}^{-1}}\left(\frac{1}{T_2} - \frac{1}{623.2 \text{ K}}\right)$$

$$T_2 = 644.0 \text{ K} = 371° \text{ C}$$

9.34   Consider the following parallel reactions

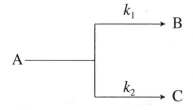

The activation energies are 45.3 kJ mol^{-1} for k_1 and 69.8 kJ mol^{-1} for k_2. If the rate constants are equal at 320 K, at what temperature will $k_1/k_2 = 2.00$?

The ratio of the rate constants is

$$\frac{k_1}{k_2} = \frac{A_1 e^{-E_{a1}/RT}}{A_2 e^{-E_{a2}/RT}}$$

$$= \frac{A_1}{A_2} e^{(E_{a2}-E_{a1})/RT} = \frac{A_1}{A_2} e^{\left(69.8\times10^3 \text{ J mol}^{-1} - 45.3\times10^3 \text{ J mol}^{-1}\right)/[(8.314\text{J K}^{-1} \text{ mol}^{-1})T]}$$

$$= \frac{A_1}{A_2} e^{2.947\times10^3 \text{ K}/T}$$

First use data at 320 K to calculate A_1/A_2:

$$\frac{k_1}{k_2} = 1.00 = \frac{A_1}{A_2} e^{2.947\times10^3 \text{ K}/320 \text{ K}}$$

$$\frac{A_1}{A_2} = 1.001 \times 10^{-4}$$

When $k_1/k_2 = 2.00$,

$$2.00 = \frac{A_1}{A_2} e^{2.947\times10^3 \text{ K}/T} = \left(1.001 \times 10^{-4}\right) e^{2.947\times10^3 \text{ K}/T}$$

$$\frac{1}{T} = \frac{1}{2.947 \times 10^3 \text{ K}} \ln \frac{2.00}{1.001 \times 10^{-4}} = 3.360 \times 10^{-3} \text{ K}^{-1}$$

$$T = 298 \text{ K}$$

9.36 The rate of the electron-exchange reaction between naphthalene ($C_{10}H_8$) and its anion radical ($C_{10}H_8^-$) is diffusion-controlled:

$$C_{10}H_8^- + C_{10}H_8 \rightleftharpoons C_{10}H_8 + C_{10}H_8^-$$

The reaction is bimolecular and second order. The rate constants are

T/K	307	299	289	273
$k/10^9 \cdot M^{-1} \cdot s^{-1}$	2.71	2.40	1.96	1.43

Calculate the values of E_a, $\Delta H^{o\ddagger}$, $\Delta S^{o\ddagger}$ and $\Delta G^{o\ddagger}$ at 307 K for the reaction. [*Hint*: Rearrange Equation 9.41 and plot $\ln (k/T)$ versus $1/T$.]

Equation 9.41 gives

$$k = \frac{k_B T}{h} e^{\Delta S^{o\ddagger}/R} e^{-\Delta H^{o\ddagger}/RT}$$

or

$$\ln \frac{k}{T} = \ln \frac{k_B}{h} + \frac{\Delta S^{o\ddagger}}{R} - \frac{\Delta H^{o\ddagger}}{RT}$$

A plot of $\ln k/T$ vs $1/T$ gives a slope of $-\Delta H^{o\ddagger}/R$ and an intercept of $\ln k_B/h + \Delta S^{o\ddagger}/R$. The data used for the plot are

10^3 K/T	3.257	3.344	3.460	3.663
$\ln \frac{k}{T}$	15.9934	15.8983	15.7298	15.4715

The best fit line has a formula of $y = -1302.0x + 20.24$. Therefore,

$$\Delta H^{\text{o}\ddagger} = -(-1302.0 \text{ K}) \left(8.314 \text{ J K}^{-1} \text{ mol}^{-1}\right)$$

$$= 1.082 \times 10^4 \text{ J mol}^{-1}$$

$$= 1.08 \times 10^4 \text{ J mol}^{-1}$$

and

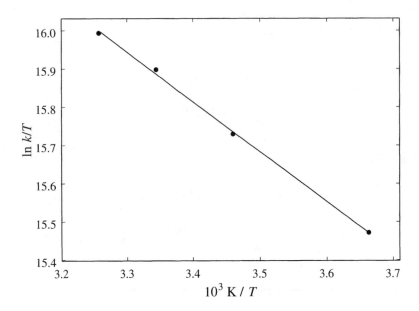

$$\Delta S^{\text{o}\ddagger} = R \left(20.24 - \ln \frac{k_{\text{B}}}{h}\right)$$

$$= \left(8.314 \text{ J K}^{-1} \text{mol}^{-1}\right) \left(20.24 - \ln \frac{1.381 \times 10^{-23}}{6.626 \times 10^{-34}}\right)$$

$$= -29.3 \text{ J K}^{-1} \text{mol}^{-1}$$

From Equation 9.43 and the discussion following it, the activation energy for this reaction, which occurs in solution (condensed phase), is

$$E_{\text{a}} = \Delta H^{\text{o}\ddagger} + RT$$

$$= 1.082 \times 10^4 \text{ J mol}^{-1} + \left(8.314 \text{ J K}^{-1} \text{mol}^{-1}\right) (307 \text{ K})$$

$$= 1.34 \times 10^4 \text{ J mol}^{-1}$$

From $\Delta H^{\text{o}\ddagger}$ and $\Delta S^{\text{o}\ddagger}$, $\Delta G^{\text{o}\ddagger}$ at 307 K is calculated.

$$\Delta G^{\text{o}\ddagger} = \Delta H^{\text{o}\ddagger} - T \Delta S^{\text{o}\ddagger}$$

$$= 1.082 \times 10^4 \text{ J mol}^{-1} - (307 \text{ K}) \left(-29.3 \text{ J K}^{-1} \text{mol}^{-1}\right)$$

$$= 1.98 \times 10^4 \text{ J mol}^{-1}$$

9.38 A person may die after drinking D_2O instead of H_2O for a prolonged period (on the order of days). Explain. Because D_2O has practically the same properties as H_2O, how would you test the presence of large quantities of the former in a victim's body?

Because of the lower zero-point energy for bonds in which D is substituted for H, there is a higher activation energy required for reactions in which this bond breaks. Thus, the rate of H^+ ion exchange is faster than that for the D^+ ion. Additionally, the dissociation constants of deuterated acids are smaller than the corresponding acid with the normal H^+ ion. These differences will affect the delicate acid-base balance in the body as well as the kinetics of biological processes, and could lead to death. A mass spectrum of a body fluid sample should reveal the presence of a larger than natural abundance of the heavier isotope of hydrogen.

9.40 Lubricating oils for watches or other mechanical objects are made of long-chain hydrocarbons. Over long-periods of time they undergo auto-oxidation to form solid polymers. The initial step in this process involves hydrogen abstraction. Suggest a chemical means for prolonging the life of these oils.

Deuterating the oils, that is, replacing the H atoms with D atoms, will slow down the rate of hydrogen abstraction.

9.42 The term *reversible* is used in both thermodynamics (see Chapter 3) and in this chapter. Does it convey the same meaning in these two instances?

The term has a different meaning in kinetics than it does in thermodynamics. In kinetics, a reversible reaction is one in which both the forward and the backward reaction occur. In thermodynamics, a reversible process is one that is in equilibrium at every point along the path connecting the initial and final states.

9.44 The equilibrium between dissolved CO_2 and carbonic acid can be represented by

$$H^+ + HCO_3^- \underset{k_{21}}{\overset{k_{12}}{\rightleftharpoons}} H_2CO_3$$

$$k_{13} \Updownarrow k_{31} \qquad\qquad k_{23} \Updownarrow k_{32}$$

$$CO_2 \qquad + \qquad H_2O$$

Show that

$$-\frac{d\,[CO_2]}{dt} = \left(k_{31} + k_{32}\right)[CO_2] - \left(k_{13} + \frac{k_{23}}{K}\right)[H^+]\,[HCO_3^-]$$

where $K = [H^+]\,[HCO_3^-]/[H_2CO_3]$.

From the equilibrium,

$$\frac{d\,[CO_2]}{dt} = k_{13}[H^+][HCO_3^-] - k_{31}[CO_2] + k_{23}[H_2CO_3] - k_{32}[CO_2]$$

Since H_2O is present in a great quantity, the effectively constant concentration, $[H_2O]$, is incorporated into the constants k_{31} and k_{32}. Rearranging the expression gives

$$\frac{d\,[CO_2]}{dt} = -\left(k_{31} + k_{32}\right)[CO_2] + k_{13}[H^+][HCO_3^-] + k_{23}[H_2CO_3] \qquad (9.44.1)$$

Since H^+ and HCO_3^- are in equilibrium with H_2CO_3, let

$$K = \frac{[H^+][HCO_3^-]}{[H_2CO_3]}$$

$$[H_2CO_3] = \frac{[H^+][HCO_3^-]}{K} \qquad (9.44.2)$$

Substitute Eq. 9.44.2 into Eq. 9.44.1,

$$\frac{d\,[CO_2]}{dt} = -\left(k_{31} + k_{32}\right)[CO_2] + k_{13}[H^+][HCO_3^-] + k_{23}\frac{[H^+][HCO_3^-]}{K}$$

$$= -\left(k_{31} + k_{32}\right)[CO_2] + \left(k_{13} + \frac{k_{23}}{K}\right)[H^+][HCO_3^-]$$

$$-\frac{d\,[CO_2]}{dt} = \left(k_{31} + k_{32}\right)[CO_2] - \left(k_{13} + \frac{k_{23}}{K}\right)[H^+][HCO_3^-]$$

9.46   In a certain industrial process involving a heterogeneous catalyst, the volume of the catalyst (in the shape of a sphere) is 10.0 cm³. **(a)** Calculate the surface area of the catalyst. **(b)** If the sphere is broken down into eight spheres, each of which has a volume of 1.25 cm³, what is the total surface area of the spheres? **(c)** Which of the two geometric configurations is the more effective catalyst? (*Hint*: The surface area of a sphere is $4\pi r^2$, where r is the radius of the sphere.)

(a) One 10.0 cm³ sphere

First calculate the radius of the sphere.

$$V = 10.0\ cm^3 = \frac{4}{3}\pi r^3$$

$$r = 1.337\ cm$$

The surface area is

$$A = 4\pi r^2 = 4\pi\,(1.337\ cm)^2 = 22.5\ cm^2$$

(b) Eight 1.25 cm³ spheres

First calculate the radius of one sphere.

$$V = 1.25 \text{ cm}^3 = \frac{4}{3}\pi r^3$$

$$r = 0.6683 \text{ cm}$$

The total surface area of the spheres is

$$A = 8\left(4\pi r^2\right) = 32\pi \ (0.6683 \text{ cm})^2 = 44.9 \text{ cm}^2$$

(c) Since a greater surface area promotes the catalyzed reaction more effectively, the eight smaller spheres are more effective than one large sphere.

9.48 At a certain elevated temperature, ammonia decomposes on the surface of tungsten metal as follows:

$$NH_3 \longrightarrow \frac{1}{2}N_2 + \frac{3}{2}H_2$$

The kinetic data are expressed as the variation of the half-life with the initial pressure of NH_3:

P/torr	264	130	59	16
$t_{1/2}$/s	456	228	102	60

(a) Determine the order of the reaction. **(b)** How does the order depend on the initial pressure? **(c)** How does the mechanism of the reaction vary with pressure?

(a) The half-life of a reaction and the initial concentration are related by

$$t_{1/2} = C\frac{1}{[A]_0^{n-1}}$$

where C is a constant. Taking the common logarithm of both sides of the equation,

$$\log t_{1/2} = \log C - (n-1)\log [A]_0$$

Since pressure is proportional to concentration at constant temperature, the above equation can also be written as

$$\log t_{1/2} = \log C' - (n-1)\log P$$

A plot of $\log t_{1/2}$ vs $\log P$ gives a slope of $n-1$. The data used for the plot are

$\log(P/\text{torr})$	2.422	2.114	1.77	1.20
$\log(t_{1/2}/\text{s})$	2.659	2.358	2.009	1.78

There are clearly two types of behavior exhibited in the graph. At pressures above 50 torr the graph appears to be a straight line, and fitting to these three points results in a best fit line with an equation of $y = 1.00x + 0.24$. Thus, $1 = -(n-1)$, or $n = 0$, and the reaction is zero-order.

Although the data are limited, it is clear that there is a change in slope below 50 torr, indicating a change in reaction order. It does appear that the limiting slope as pressure approaches zero is itself zero. Thus, $0 = -(n-1)$, or $n = 1$, and the limiting behavior is that of a first-order reaction.

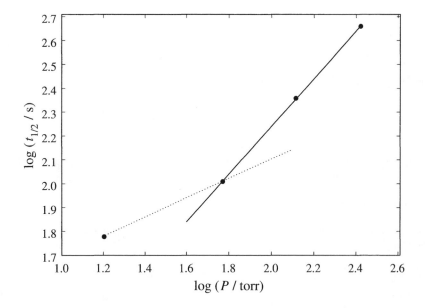

(b) As discovered in part **(a)**, the reaction is first order at low pressures and zero order at pressures above 50 torr.

(c) The mechanism is actually the same at all pressures considered. At low pressures, the fraction of the tungsten surface covered is proportional to the pressure of NH_3, so the rate of decomposition will have a first order dependence on ammonia pressure. As discussed in Problem 9.16, however, at increased pressures, all the catalytic sites are occupied by, in this case, NH_3 molecules and the rate becomes independent of the ammonia pressure and zero order in NH_3.

9.50 The reaction $X \rightarrow Y$ has a reaction enthalpy of -64 kJ mol^{-1} and an activation energy of 22 kJ mol^{-1}. What is the activation energy for the $Y \rightarrow X$ reaction?

Referring to the figure, the activation energy for the reverse reaction, $Y \rightarrow X$, is seen to be 22 kJ mol^{-1} + 64 kJ mol^{-1} = 86 kJ mol^{-1}.

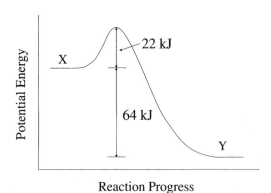

9.52 As a result of being exposed to the radiation released during the Chernobyl nuclear accident, a person had a level of iodine-131 in his body equal to 7.4 mCi (1 mCi = 1 × 10^{-3} Ci). Calculate the number of atoms of I-131 to which this radioactivity corresponds. Why were people who lived close to the nuclear reactor site urged to take large amounts of potassium iodide after the accident?

For nuclear decay

$$\text{rate} = \lambda N = \frac{\ln 2}{t_{1/2}} N$$

where N is the number of radioactive atoms. The rate is given by the activity,

$$\text{rate} = (7.4 \text{ mCi}) \left(\frac{1 \text{ Ci}}{1000 \text{ mCi}} \right) \left(\frac{3.70 \times 10^{10} \text{ s}^{-1}}{1 \text{ Ci}} \right)$$

$$= 2.74 \times 10^8 \text{ s}^{-1}$$

where the activity corresponding to 1 Ci is found in Problem 9.49. The half-life for I-131 is found in Table 9.1.

$$N = \left(\frac{t_{1/2}}{\ln 2} \right) \times \text{rate}$$

$$= \left(\frac{8.05 \text{ d}}{\ln 2} \right) \left(\frac{24 \text{ h}}{1 \text{ d}} \right) \left(\frac{3600 \text{ s}}{1 \text{ h}} \right) \left(2.74 \times 10^8 \text{ s}^{-1} \right)$$

$$= 2.7 \times 10^{14}$$

The human body concentrates iodine in the thyroid gland. Large doses of (non-radioactive) KI will displace the radioactive iodine from the thyroid and allow its excretion from the body.

9.54 The bromination of acetone is acid-catalyzed:

$$CH_3COCH_3 + Br_2 \xrightarrow{H^+} CH_3COCH_2Br + H^+ + Br^-$$

The rate of disappearance of bromine was measured for several different concentrations of acetone, bromine, and H^+ ions at a certain temperature:

	$[CH_3COCH_3]$/M	$[Br_2]$/M	$[H^+]$/M	Rate of Disappearance of Br_2 /$M \cdot s^{-1}$
(1)	0.30	0.050	0.050	5.7×10^{-5}
(2)	0.30	0.10	0.050	5.7×10^{-5}
(3)	0.30	0.050	0.1	1.2×10^{-4}
(4)	0.40	0.050	0.2	3.1×10^{-4}
(5)	0.40	0.050	0.050	7.6×10^{-5}

(a) What is the rate law for the reaction? **(b)** Determine the rate constant. **(c)** The following mechanism has been proposed for the reaction:

$$H_3C-\overset{\overset{\displaystyle O}{\|}}{C}-CH_3 + H_3O^+ \rightleftharpoons H_3C-\overset{\overset{\displaystyle OH^+}{\|}}{C}-CH_3 \quad \text{(fast equilibrium)}$$

$$H_3C-\overset{\overset{\displaystyle OH^+}{\|}}{C}-CH_3 + H_2O \longrightarrow H_3C-\overset{\overset{\displaystyle OH}{|}}{C}=CH_2 + H_3O^+ \quad \text{(slow)}$$

$$H_3C-\overset{\overset{\displaystyle OH}{|}}{C}=CH_2 + Br_2 \longrightarrow H_3C-\overset{\overset{\displaystyle O}{\|}}{C}-CH_2Br + HBr \quad \text{(fast)}$$

Show that the rate law deduced from the mechanism is consistent with that shown in (a).

(a) Comparing Experiment (1) and Experiment (5), the concentrations of Br_2 and H^+ are constant and the concentration of CH_3COCH_3 has increased 1.33 times. The rate has also increased 1.33 times. Therefore, the rate is directly proportional to the concentration of CH_3COCH_3. That is, the reaction is first order in CH_3COCH_3.

Comparing Experiment (1) and Experiment (2), the concentrations of CH_3COCH_3 and H^+ are constant and the concentration of Br_2 has increased 2.0 times but the rate has not changed. Therefore, the rate is independent of the concentration of Br_2. That is, the reaction is zeroth order in Br_2.

Comparing Experiment (1) and Experiment (3), the concentrations of CH_3COCH_3 and Br_2 are constant and the concentration of H^+ has increased 2.0 times. The rate has increased 2.1 times. Therefore, the rate is directly proportional to the concentration of H^+. That is, the reaction is first order in H^+.

Therefore, the rate law is

$$\text{Rate} = k[CH_3COCH_3][H^+]$$

(b) The rate constant can be calculated using data from any experiment. Using Exp. 1,

$$\text{Rate} = 5.7 \times 10^{-5}\ M\ s^{-1} = k\ (0.30\ M)\ (0.050\ M)$$

$$k = 3.8 \times 10^{-3}\ M^{-1} s^{-1}$$

(c) Since step 2 is the slow step,

$$\text{Rate} = k_2[CH_3COHCH_3^+][H_2O] \tag{9.54.1}$$

The intermediate, $CH_3COHCH_3^+$ can be written in terms of the reactants using the first fast equilibrium step.

$$\text{Forward rate of step 1} = \text{Reverse rate of step 1}$$

$$k_1[CH_3COCH_3][H_3O^+] = k_{-1}[CH_3COHCH_3^+][H_2O]$$

$$[CH_3COHCH_3^+][H_2O] = \frac{k_1}{k_{-1}}[CH_3COCH_3][H_3O^+] \tag{9.54.2}$$

Substituting Eq. 9.54.2 into Eq. 9.54.1, the rate law becomes

$$\text{Rate} = \frac{k_2 k_1}{k_{-1}}[CH_3COCH_3][H_3O^+] = \frac{k_2 k_1}{k_{-1}}[CH_3COCH_3][H^+]$$

which has the same form as that shown in **(a)**.

9.56 For the cyclic reactions shown on p. 329, show that $k_{-1}k_2k_3 = k_1k_{-2}k_{-3}$.

According to the reactions,

$$k_2[A] = k_{-2}[B]$$

$$k_{-1}[C] = k_1[A]$$

$$k_3[B] = k_{-3}[C]$$

Therefore,

$$k_2[A]k_{-1}[C]k_3[B] = k_{-2}[B]k_1[A]k_{-3}[C]$$

$$k_{-1}k_2k_3 = k_1k_{-2}k_{-3}$$

9.58 Sucrose ($C_{12}H_{22}O_{11}$), commonly called table sugar, undergoes hydrolysis (reaction with water) to produce fructose ($C_6H_{12}O_6$) and glucose ($C_6H_{12}O_6$):

$$C_{12}H_{22}O_{11} \quad + \quad H_2O \quad \longrightarrow \quad \underset{\text{fructose}}{C_6H_{12}O_6} \quad + \quad \underset{\text{glucose}}{C_6H_{12}O_6}$$

This reaction has particular significance in the candy industry. First, fructose is sweeter than sucrose. Second, a mixture of fructose and glucose, called *invert* sugar, does not crystallize, so candy made with this combination is chewier and not brittle as crystalline sucrose is. Sucrose is dextrorotatory (+), whereas the mixture of glucose and fructose resulting from inversion in levorotatory (−). Thus, a decrease in the concentration of sucrose will be accompanied by a proportional decrease in the optical rotation. **(a)** From the following kinetic data, show that the reaction is first order, and determine the rate constant.

time/min	0	7.20	18.0	27.0	∞
optical rotation (α)	+24.08°	+21.40°	+17.73°	+15.01°	−10.73°

(b) Explain why the rate law does not include [H_2O] even though water is a reactant.

(a) The total change in rotation (from $t = 0$ to $t = \infty$), given by $(\alpha_0 - \alpha_\infty)$, will be proportional to the decrease in concentration of sucrose by that time. Therefore, the concentration of sucrose remaining at time t will be proportional to

$$(\alpha_0 - \alpha_\infty) - (\alpha_0 - \alpha_t) = \alpha_t - \alpha_\infty$$

If the reaction is first order, then

$$\ln \frac{[\text{sucrose}]}{[\text{sucrose}]_0} = \ln \frac{\alpha_t - \alpha_\infty}{\alpha_0 - \alpha_\infty} = -kt$$

A plot of $\left[\ln(\alpha_t - \alpha_\infty)/(\alpha_0 - \alpha_\infty)\right]$ vs t should give a straight line with a slope of $-k$. The plot uses the following data:

time/min	0	7.20	18.0	27.0
$\ln\left(\alpha_t - \alpha_\infty\right)/\left(\alpha_0 - \alpha_\infty\right)$	0	-8.0115×10^{-2}	-0.20141	-0.30186

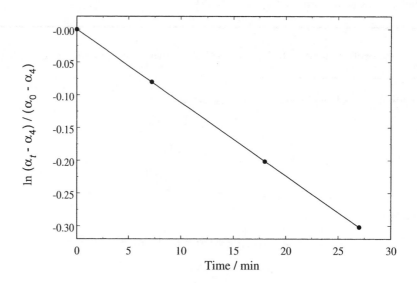

The plot indeed results in a line described by $y = -1.112 \times 10^{-2}x + 1.5 \times 10^{-4}$. Therefore, the rate constant is 1.11×10^{-2} min^{-1}.

(b) The rate law does not include [H$_2$O] because water is present in very high concentration (ca. 55.5 M) and this is a pseudo first-order reaction.

9.60   Under certain conditions the gas-phase decomposition of ozone is found to be second order in O$_3$ and inhibited by molecular oxygen. Apply the steady-state approximation to the following mechanism to show that the rate law is consistent with the experimental observations:

$$O_3 \underset{k_{-1}}{\overset{k_1}{\rightleftharpoons}} O_2 + O$$

$$O + O_3 \overset{k_2}{\longrightarrow} 2O_2$$

State any assumption made in the derivation.

The rate of decomposition of O$_3$ is

$$-\frac{d\,[O_3]}{dt} = k_1[O_3] - k_{-1}[O_2][O] + k_2[O][O_3]$$

$$= k_1[O_3] + \left(k_2[O_3] - k_{-1}[O_2]\right)[O] \tag{9.60.1}$$

Apply the steady state approximation to O:

$$\frac{d\,[O]}{dt} = k_1[O_3] - k_{-1}[O_2][O] - k_2[O][O_3] = 0$$

$$[O] = \frac{k_1[O_3]}{k_{-1}[O_2] + k_2[O_3]} \tag{9.60.2}$$

Substitute Eq. 9.60.2 into Eq. 9.60.1:

$$-\frac{d\,[O_3]}{dt} = k_1[O_3] + \left(k_2[O_3] - k_{-1}[O_2]\right) \frac{k_1[O_3]}{k_{-1}[O_2] + k_2[O_3]}$$

$$= \frac{k_1 k_{-1}[O_3][O_2] + k_1 k_2[O_3]^2}{k_{-1}[O_2] + k_2[O_3]} + \frac{k_1 k_2[O_3]^2}{k_{-1}[O_2] + k_2[O_3]} - \frac{k_1 k_{-1}[O_3][O_2]}{k_{-1}[O_2] + k_2[O_3]}$$

$$= \frac{2k_1 k_2[O_3]^2}{k_{-1}[O_2] + k_2[O_3]} \qquad (9.60.3)$$

If the rate of the second step is assumed to be much slower than the rate of the reverse reaction for the first step, then

$$k_2[O][O_3] \ll k_{-1}[O_2][O]$$

$$k_2[O_3] \ll k_{-1}[O_2]$$

Equation 9.60.3 becomes

$$-\frac{d\,[O_3]}{dt} = \frac{2k_1 k_2[O_3]^2}{k_{-1}[O_2]}$$

Since the rate for the reaction $2O_3 \rightarrow 3O_2$ is $-\frac{1}{2}\frac{d\,[O_3]}{dt}$, the rate law predicted by this mechanism is

$$\text{Rate} = -\frac{1}{2}\frac{d\,[O_3]}{dt} = \frac{k_1 k_2[O_3]^2}{k_{-1}[O_2]}$$

which is consistent with experimental observations.

9.62 A reaction $X + Y \rightarrow Z$ proceeds by two different mechanisms, one of which is pH dependent. The rate law for the reaction is

$$\frac{d\,[Z]}{dt} = k_1\,[X] + k_2\,[Y]\left[H^+\right]$$

At pH = 3.4 the rates of the two reactions are equal. What is the ratio k_1/k_2? Assume $[X] = [Y]$.

When the rates of the two reactions are equal,

$$k_1\,[X] = k_2\,[Y]\left[H^+\right]$$

Rearrange the equation and set $[X] = [Y]$ and $\left[H^+\right] = 10^{-3.4}$

$$\frac{k_1}{k_2} = \frac{[Y]\left[H^+\right]}{[X]} = 10^{-3.4}\,M$$

$$= 4 \times 10^{-4}\,M$$

9.64 The Polish-American physicist Roman Smoluchowski showed that the rate constant for a diffusion-controlled reaction between molecules A and B, k_D, is given by

$$k_D = 4\pi N_A \left(D_A + D_B\right) r_{AB}$$

where r_{AB} is the distance (in cm) between the molecules and D_A and D_B are their diffusion coefficients. Calculate k_D, given that $D_A = 6.0 \times 10^{-5}\ cm^2\ s^{-1}$, $D_B = 2.5 \times 10^{-5}\ cm^2\ s^{-1}$ at $20°\ C$, and $r_{AB} = 1.0 \times 10^{-8}\ cm$.

$$k_D = 4\pi N_A \left(D_A + D_B\right) r_{AB}$$

$$= 4\pi \left(6.022 \times 10^{23}\ mol^{-1}\right) \left(6.0 \times 10^{-5}\ cm^2\ s^{-1} + 2.5 \times 10^{-5}\ cm^2\ s^{-1}\right) \left(1.0 \times 10^{-8}\ cm\right)$$

$$= \left(6.43 \times 10^{12}\ cm^3\ s^{-1}\ mol^{-1}\right) \left(\frac{1\ L}{1000\ cm^3}\right)$$

$$= 6.43 \times 10^9\ L\ s^{-1}\ mol^{-1} = 6.4 \times 10^9\ M^{-1}\ s^{-1}$$

Note that a truly diffusion-controlled reaction does not have an energy of activation.

CHAPTER 10

Enzyme Kinetics

PROBLEMS AND SOLUTIONS

10.2 Measurements of a certain enzyme-catalyzed reaction give $k_1 = 8 \times 10^6\ M^{-1}\,s^{-1}$, $k_{-1} = 7 \times 10^4\ s^{-1}$, and $k_2 = 3 \times 10^3\ s^{-1}$. Does the enzyme–substrate binding follow the equilibrium or steady-state scheme?

The dissociation constant, K_S, and the Michaelis constant, K_M, must be compared.

$$\begin{aligned} K_S &= \frac{k_{-1}}{k_1} \\ &= \frac{7 \times 10^4\ s^{-1}}{8 \times 10^6\ M^{-1}\,s^{-1}} \\ &= 9 \times 10^{-3}\ M \end{aligned}$$

and

$$\begin{aligned} K_M &= \frac{k_{-1} + k_2}{k_1} \\ &= \frac{7 \times 10^4\ s^{-1} + 3 \times 10^3\ s^{-1}}{8 \times 10^6\ M^{-1}\,s^{-1}} \\ &= 9 \times 10^{-3}\ M \end{aligned}$$

Within the precision of the measurements, the two constants are equal. Thus, the binding follows the equilibrium scheme. That is, k_{-1} is sufficiently greater than k_2 so that the binding reaches equilibrium.

10.4 Derive the following equation from Equation 10.10,

$$\frac{v_0}{[S]} = \frac{V_{max}}{K_M} - \frac{v_0}{K_M}$$

and show how you would obtain values of K_M and V_{max} graphically from this equation.

Starting with Equation 10.10, multiply both sides by $K_M + [S]$, then divide by $K_M[S]$ and rearrange.

$$v_0 = \frac{V_{max}[S]}{K_M + [S]}$$

$$v_0 K_M + v_0[S] = V_{max}[S]$$

$$v_0 K_M = V_{max}[S] - v_0[S]$$

$$\frac{v_0}{[S]} = \frac{V_{max}}{K_M} - \frac{v_0}{K_M}$$

Thus, a plot of $v_0/[S]$ vs. v_0 will have a slope of $-1/K_M$ and a y-intercept of V_{max}/K_M. The same data, however, when plotted in a Eadie–Hofstee plot, v_0 vs. $v_0/[S]$, gives more straightforward results (see Problem 10.6).

10.6 The hydrolysis of N-glutaryl-L-phenylalanine-p-nitroanilide (GPNA) to p-nitroaniline and N-glutaryl-L-phenylalanine is catalyzed by α-chymotrypsin. The following data are obtained:

$[S]/10^{-4}\ M$	2.5	5.0	10.0	15.0
$v_0/10^{-6}\ M \cdot min^{-1}$	2.2	3.8	5.9	7.1

where $[S] = [GPNA]$. Assuming Michaelis–Menten kinetics, calculate the values of V_{max}, K_M, and k_2 using the Lineweaver–Burk plot. Another way to treat the data is to plot v_0 versus $v_0/[S]$, which is the Eadie–Hofstee plot. Calculate the values of V_{max}, K_M, and k_2 from the Eadie–Hofstee treatment, given that $[E]_0 = 4.0 \times 10^{-6}\ M$. [*Source*: J. A. Hurlbut, T. N. Ball, H. C. Pound, and J. L. Graves, *J. Chem. Educ.* **50**, 149 (1973).]

For the Lineweaver–Burk plot, the following data are needed.

$(1/[S])/10^3 \cdot M^{-1}$	4.00	2.00	1.00	0.667
$(1/v_0)/10^5 \cdot M^{-1} \cdot min$	4.55	2.63	1.69	1.41

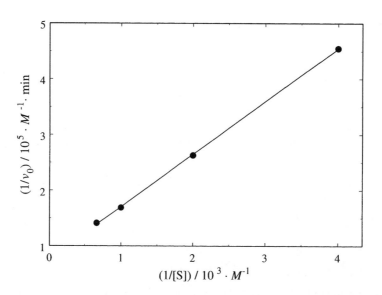

The best-fit line to the data has an equation of $y = 94.6x + 7.56 \times 10^4$. The intercept of a Lineweaver–Burk plot is $1/V_{max}$ giving

$$V_{max} = \frac{1}{7.56 \times 10^4 \, M^{-1} \, min}$$

$$= 1.32 \times 10^{-5} \, M \, min^{-1}$$

$$= 1.3 \times 10^{-5} \, M \, min^{-1}$$

The slope is K_M / V_{max} so that

$$K_M = (94.6 \, min)(1.32 \times 10^{-5} \, M \, min^{-1})$$

$$= 1.2 \times 10^{-3} \, M$$

Finally,

$$k_2 = \frac{V_{max}}{[E]_0}$$

$$= \frac{1.32 \times 10^{-5} \, M \, min^{-1}}{4.0 \times 10^{-6} \, M}$$

$$= 3.3 \, min^{-1}$$

The Eadie–Hofstee plot uses the following data,

$(v_0/[S])/10^{-3} \cdot min^{-1}$	8.80	7.60	5.90	4.73
$v_0 /10^{-6} \cdot M \cdot min^{-1}$	2.2	3.8	5.9	7.1

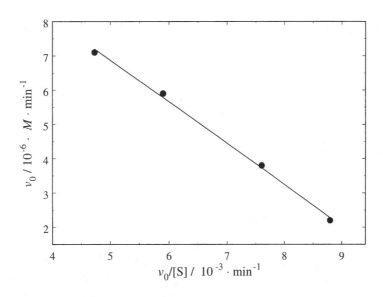

The best-fit line to the data has an equation of $y = -1.21 \times 10^{-3} x + 1.29 \times 10^{-5}$. In a Eadie–Hofstee plot the slope is $-K_M$ and the y-intercept is V_{max}. Thus, $V_{max} = 1.3 \times 10^{-5} \, M \, min^{-1}$ and $K_M = 1.2 \times 10^{-3} \, M$. $k_2 = 3.3 \, min^{-1}$ is found as above. These are the same values as found from the Lineweaver–Burk plot, which given good data is as expected. The two plots weight the data differently, so that the values determined may be different depending on the quality of the data.

10.8   The hydrolysis of urea,

$$(NH_2)_2CO + H_2O \rightarrow 2NH_3 + CO_2$$

has been studied by many researchers. At 100°C, the (pseudo) first-order rate constant is 4.2×10^{-5} s^{-1}. The reaction is catalyzed by the enzyme urease, which at 21°C has a rate constant of 3×10^4 s^{-1}. If the enthalpies of activation for the uncatalyzed and catalyzed reactions are 134 kJ mol^{-1} and 43.9 kJ mol^{-1}, respectively, **(a)** calculate the temperature at which the nonenzymatic hydrolysis of urea would proceed at the same rate as the enzymatic hydrolysis at 21°C; **(b)** calculate the lowering of $\Delta G^{\ddagger}$ due to urease; and **(c)** comment on the sign of $\Delta S^{\ddagger}$. Assume that $\Delta H^{\ddagger} = E_{\rm a}$ and that $\Delta H^{\ddagger}$ and $\Delta S^{\ddagger}$ are independent of temperature.

(a) The Arrhenius equation relates reaction rate and activation energy via $k = Ae^{-E_{\rm a}/RT}$. Requiring that the rates of the catalyzed and uncatalyzed reactions be equal at their respective temperatures then means (assuming A to be the same for both the catalyzed and uncatalyzed reactions),

$$k_{\rm cat} = k_{\rm uncat}$$

$$Ae^{-E_{\rm a}^{\rm cat}/RT_1} = Ae^{-E_{\rm a}^{\rm uncat}/RT_2}$$

$$\frac{E_{\rm a}^{\rm cat}}{T_1} = \frac{E_{\rm a}^{\rm uncat}}{T_2}$$

taking $E_{\rm a} \approx \Delta H^{\ddagger}$,

$$\frac{43.9 \times 10^3 \text{ J mol}^{-1}}{294.15 \text{ K}} = \frac{134 \times 10^3 \text{ J mol}^{-1}}{T_2}$$

$$T_2 = 898 \text{ K}$$

At this temperature the solvent would be vaporized and the urea thermally decomposed, so that it is in fact impossible to achieve the enzymatic rate without the catalyst.

(b) From Equation 9.40, $k = \frac{k_{\rm B}T}{h}e^{-\Delta G^{\ddagger}/RT}$, or $\Delta G^{\ddagger} = -RT \ln \frac{kh}{k_{\rm B}T}$.

For the uncatalyzed reaction at 373 K,

$$\Delta G^{\ddagger} = -\left(8.314 \text{ J K}^{-1}\text{mol}^{-1}\right)(373.15 \text{ K}) \ln \frac{\left(4.2 \times 10^{-5} \text{ s}^{-1}\right)\left(6.626 \times 10^{-34} \text{ J s}\right)}{\left(1.381 \times 10^{-23} \text{ J K}^{-1}\right)(373.15 \text{ K})}$$

$$= 1.234 \times 10^5 \text{ J mol}^{-1}$$

For the catalyzed reaction at 294 K,

$$\Delta G^{\ddagger} = -\left(8.314 \text{ J K}^{-1}\text{mol}^{-1}\right)(294.15 \text{ K}) \ln \frac{\left(3 \times 10^4 \text{ s}^{-1}\right)\left(6.626 \times 10^{-34} \text{ J s}\right)}{\left(1.381 \times 10^{-23} \text{ J K}^{-1}\right)(294.15 \text{ K})}$$

$$= 4.70 \times 10^4 \text{ J mol}^{-1}$$

Thus, $\Delta G^{\ddagger}$ is lowered by 1.234×10^5 J mol$^{-1} - 4.70 \times 10^4$ J mol$^{-1} = 7.64 \times 10^4$ J mol^{-1}, although the comparison is being made at two different temperatures.

(c) Since $\Delta G^{\ddagger} = \Delta H^{\ddagger} - T\Delta S^{\ddagger}$, $\Delta S^{\ddagger} = \left(\Delta H^{\ddagger} - \Delta G^{\ddagger}\right)/T$.

For the uncatalyzed reaction,

$$\Delta S^{\ddagger} = \frac{134 \times 10^3 \text{ J mol}^{-1} - 1.234 \times 10^5 \text{ J mol}^{-1}}{373.15 \text{ K}} = 28.4 \text{ J K}^{-1}\text{mol}^{-1}$$

There is an increase in entropy upon approaching the transition state as would be expected in a case where a single molecule is breaking apart in two or more fragments in the transition state.

For the catalyzed reaction,

$$\Delta S^{\ddagger} = \frac{43.9 \times 10^3 \text{ J mol}^{-1} - 4.70 \times 10^4 \text{ J mol}^{-1}}{294.15 \text{ K}} = -11 \text{ J K}^{-1} \text{ mol}^{-1}$$

Here there is a decrease in entropy upon approaching the transition state, since the rate determining step now involves the binding of two molecules, enzyme and substrate.

10.10   Silver ions are known to react with the sulfhydryl groups of proteins and therefore can inhibit the action of certain enzymes. In one reaction, 0.0075 g of $AgNO_3$ is needed to completely inactivate a 5-mL enzyme solution. Estimate the molar mass of the enzyme. Explain why the molar mass obtained represents the minimum value. The concentration of the enzyme solution is such that 1 mL of the solution contains 75 mg of the enzyme.

The number of moles of $AgNO_3$ used to inactivate the enzyme is

$$\frac{7.5 \times 10^{-3} \text{ g}}{169.9 \text{ g mol}^{-1}} = 4.41 \times 10^{-5} \text{ mol}$$

Assuming 1:1 binding between the silver ions and the protein, this is also the number of moles of enzyme present in the 5 mL solution which contains $(5 \text{ mL}) \left(75 \times 10^{-3} \text{ g mL}^{-1}\right) = 0.375 \text{ g}$ of enzyme. Thus, the molar mass of the enzyme is

$$\frac{0.375 \text{ g}}{4.41 \times 10^{-5} \text{ mol}} = 8.5 \times 10^3 \text{ g mol}^{-1}$$

This is the minimum value for the molar mass because of the assumption of 1:1 binding. If there were more than one Ag^+ ion binding site per enzyme, there would be fewer moles of enzyme present leading to a larger value for the molar mass.

10.12   An enzyme has a K_M value of 2.8×10^{-5} M and a V_{max} value of 53 μM min^{-1}. Calculate the value of v_0 if [S] = 3.7×10^{-4} M and [I] = 4.8×10^{-4} M for (a) a competitive inhibitor, (b) a noncompetitive inhibitor, and (c) an uncompetitive inhibitor. ($K_I = 1.7 \times 10^{-5}$ M for all three cases.)

(a) For a competitive inhibitor, from Equation 10.17,

$$v_0 = \frac{V_{max}[S]}{K_M \left(1 + \frac{[I]}{K_I}\right) + [S]}$$

$$= \frac{\left(53 \ \mu M \ \text{min}^{-1}\right) \left(3.7 \times 10^{-4} \ M\right)}{\left(2.8 \times 10^{-5} \ M\right) \left(1 + \frac{4.8 \times 10^{-4} \ M}{1.7 \times 10^{-5} \ M}\right) + 3.7 \times 10^{-4} \ M}$$

$$= 16.5 \ \mu M \ \text{min}^{-1}$$

$$= 16 \ \mu M \ \text{min}^{-1}$$

(b) For a noncompetitive inhibitor, Equation 10.19 gives,

$$v_0 = \frac{\frac{V_{max}}{\left(1+\frac{[I]}{K_I}\right)}[S]}{K_M + [S]}$$

$$= \frac{\frac{53\ \mu M\ min^{-1}}{\left(1+\frac{4.8\times 10^{-4}\ M}{1.7\times 10^{-5}\ M}\right)}\left(3.7\times 10^{-4}\ M\right)}{2.8\times 10^{-5}\ M + 3.7\times 10^{-4}\ M}$$

$$= 1.69\ \mu M\ min^{-1}$$

$$= 1.7\ \mu M\ min^{-1}$$

(c) For an uncompetitive inhibitor, Equation 10.22 is appropriate,

$$v_0 = \frac{\frac{V_{max}}{\left(1+\frac{[I]}{K_I}\right)}[S]}{\frac{K_M}{\left(1+\frac{[I]}{K_I}\right)} + [S]}$$

$$= \frac{\frac{53\ \mu M\ min^{-1}}{\left(1+\frac{4.8\times 10^{-4}\ M}{1.7\times 10^{-5}\ M}\right)}\left(3.7\times 10^{-4}\ M\right)}{\frac{2.8\times 10^{-5}\ M}{\left(1+\frac{4.8\times 10^{-4}\ M}{1.7\times 10^{-5}\ M}\right)} + 3.7\times 10^{-4}\ M}$$

$$= 1.81\ \mu M\ min^{-1}$$

$$= 1.8\ \mu M\ min^{-1}$$

10.14   An enzyme-catalyzed reaction ($K_M = 2.7 \times 10^{-3}\ M$) is inhibited by a competitive inhibitor I ($K_I = 3.1 \times 10^{-5}\ M$). Suppose that the substrate concentration is $3.6 \times 10^{-4}\ M$. How much of the inhibitor is needed for 65% inhibition? How much does the substrate concentration have to be increased to reduce the inhibition to 25%?

Expressions for the initial rate in the absence and presence of a competitive inhibitor are given by Equations 10.10 and 10.17, respectively. Dividing the former by the latter gives

$$\frac{v_0}{(v_0)_{inhibition}} = \frac{K_M\left(1+\frac{[I]}{K_I}\right) + [S]}{K_M + [S]}$$

$$= 1 + \frac{K_M[I]}{\left(K_M + [S]\right)K_I}$$

This can be solved for [I],

$$[I] = K_I\left(\frac{v_0}{(v_0)_{inhibition}} - 1\right)\left(1 + \frac{[S]}{K_M}\right)$$

It can also be solved for [S],

$$[S] = K_M \left(\frac{[I]}{K_I \left(\frac{v_0}{(v_0)_{inhibition}} - 1 \right)} - 1 \right)$$

The expression for [I] is used in answering the first part of the question. For 65% inhibition, $(v_0)_{inhibition} = (1 - 0.65)v_0 = 0.35v_0$, and

$$[I] = \left(3.1 \times 10^{-5}\, M\right) \left(\frac{1}{0.35} - 1 \right) \left(1 + \frac{3.6 \times 10^{-4}\, M}{2.7 \times 10^{-3}\, M} \right) = 6.52 \times 10^{-5}\, M = 6.5 \times 10^{-5}\, M$$

To reduce the inhibition to 25%, where $(v_0)_{inhibition} = 0.75v_0$, at this concentration of inhibitor, use the expression for [S] to find the required substrate concentration.

$$[S] = \left(2.7 \times 10^{-3}\, M\right) \left[\frac{6.52 \times 10^{-5}\, M}{\left(3.1 \times 10^{-5}\, M\right)\left(\frac{1}{0.75} - 1\right)} - 1 \right] = 1.4 \times 10^{-2}\, M$$

10.16 Derive Equation 10.22.

Mass balance on the enzyme concentration gives $[E]_0 = [E] + [ES] + [ESI]$, or solving for [E] and using $K_I = \frac{[ES][I]}{[ESI]}$,

$$[E] = [E]_0 - [ES] - \frac{[ES][I]}{K_I} = [E]_0 - [ES]\left(1 + \frac{[I]}{K_I}\right)$$

Next, the steady-state approximation is applied to ES, although since the equilibrium $ES + I \rightleftharpoons ESI$ is so rapidly established, it does not affect the rates of appearance and disappearance of ES.

$$\frac{d[ES]}{dt} = 0 = k_1[E][S] - k_{-1}[ES] - k_2[ES]$$

$$= k_1[E]_0[S] - k_1[ES]\left(1 + \frac{[I]}{K_I}\right)[S] - k_{-1}[ES] - k_2[ES]$$

$$= k_1[E]_0[S] - \left[k_1\left(1 + \frac{[I]}{K_I}\right)[S] + k_{-1} + k_2\right][ES]$$

where the result from the mass balance relation has been used. This last equation is solved for [ES], using $K_M = \frac{k_{-1}+k_2}{k_1}$

$$[ES] = \frac{k_1[E]_0[S]}{k_1\left(1 + \frac{[I]}{K_I}\right)[S] + k_{-1} + k_2}$$

$$= \frac{k_1[E]_0[S]}{k_1\left(1 + \frac{[I]}{K_I}\right)[S] + k_1 K_M}$$

$$= \frac{[E]_0[S]}{\left(1 + \frac{[I]}{K_I}\right)[S] + K_M}$$

Finally, the rate of reaction is found using this expression for [ES]

$$v_0 = k_2[\text{ES}]$$

$$= \frac{k_2[\text{E}]_0[\text{S}]}{\left(1 + \frac{[\text{I}]}{K_\text{I}}\right)[\text{S}] + K_\text{M}}$$

$$= \frac{V_{\text{max}}[\text{S}]}{\left(1 + \frac{[\text{I}]}{K_\text{I}}\right)[\text{S}] + K_\text{M}}$$

$$= \frac{\frac{V_{\text{max}}}{\left(1 + \frac{[\text{I}]}{K_\text{I}}\right)}[\text{S}]}{[\text{S}] + \frac{K_\text{M}}{\left(1 + \frac{[\text{I}]}{K_\text{I}}\right)}}$$

10.18 (a) What is the physiological significance of cooperative O_2 binding by hemoglobin? Why is O_2 binding by myoglobin not cooperative? **(b)** Compare the concerted model with the sequential model for the binding of oxygen with hemoglobin.

(a) Cooperative O_2 binding enables hemoglobin to be a more efficient oxygen transporter than myoglobin. As discussed in detail in Section 10.6 of the text, nearly twice as much oxygen is delivered to the tissues than would be if O_2 binding to hemoglobin were not cooperative.

(b) The concerted model, with it's "all-or-none" limitation on the relaxed and tense forms of the four subunits in hemoglobin does not allow for the existence of mixed forms with some subunits in one form and the rest in the other. Although not relevant for the binding of oxygen with hemoglobin, the concerted model is unable to account for negative homotropic cooperativity. Nevertheless, it does allow the characterization of the allotropic behavior of hemoglobin (and enzymes) in terms of just three equilibrium constants.

The sequential model does allow for the exisistence of mixed forms of the subunits comprising the oligomer, since the binding of substrate (O_2 in the case of hemoglobin) affects only the conformation of the subunit bound to substrate. The actual mechanism of oxygen binding to hemoglobin is more complex than the limiting cases presented by the two models. The sequential model, however, does have the advantage of being able to account for negative homotropic cooperativity in enzymes displaying such behavior.

10.20 Competitive inhibitors, when present in small amounts, often act as activators to allosteric enzymes. Why?

Competitive inhibitors bind at the same active site as normal substrate. Due to the competitive nature of the binding, the presence of the inhibitor enhances the affinity of the enzyme for normal substrate.

10.22 What is the effect of each of the following actions on oxygen affinity of adult hemoglobin (Hb A) *in vitro*? **(a)** Increase pH, **(b)** increase partial pressure of CO_2, **(c)** decrease [BPG], **(d)** dissociate the tetramer into monomers, and **(e)** oxidize Fe(II) to Fe(III).

(a) Affinity increases,

(b) affinity decreases,

(c) affinity increases,

(d) affinity increases,

(e) affinity decreases.

10.24 When deoxyhemoglobin crystals are exposed to oxygen, they shatter. On the other hand, deoxymyoglobin crystals are unaffected by oxygen. Explain.

When oxygen binds to deoxyhemoglobin, the protein undergoes a conformational change. Since a crystal does not possess much flexibility, the strain caused by the molecular motion breaks the crystal. Myoglobin does not exhibit cooperativity, hence there is no conformational change upon oxygen binding, and the crystal remains intact.

10.26 The discovery in the 1980s that certain RNA molecules (the ribozymes) can act as enzymes was a surprise to many chemists. Why?

Prior to the discovery of ribozymes, all known enzymes were proteins with an immense array of varied and complex structures. RNA's on the other hand, all have relatively simple structures.

10.28 Referring to the concerted model discussed on p. 390, show that $K_1 = cL_0$.

The dissociation of the tense state with one O_2 bound (T_1) to free O_2 and the tense state with no O_2 bound (T_0) is represented by

$$T_1O_2 \rightleftharpoons T_0 + O_2$$

and has dissociation constant

$$K_T = \frac{[T_0][O_2]}{[T_1O_2]}$$

Likewise, for the relaxed form

$$R_1O_2 \rightleftharpoons R_0 + O_2$$

with dissociation constant

$$K_R = \frac{[R_0][O_2]}{[R_1O_2]}$$

Solving each of these equations for the bound form and dividing leads to

$$K_1 = \frac{[T_1O_2]}{[R_1O_2]}$$

$$= \frac{\frac{[T_0][O_2]}{K_T}}{\frac{[R_0][O_2]}{K_R}}$$

$$= \frac{[T_0]}{[R_0]}\frac{K_R}{K_T}$$

$$= cL_0$$

A similar procedure leads to

$$K_2 = c^2 L_0$$

$$K_3 = c^3 L_0$$

$$K_4 = c^4 L_0$$

10.30 (a) Comment on the following data obtained for an enzyme-catalyzed reaction (no calculations are needed):

$t/°C$	10	15	20	25	30	35	40	45
V_{max} (arbitrary units)	1.0	1.7	2.3	2.6	3.2	4.0	2.6	0.2

(b) Referring to Equation 10.8, under what conditions will an Arrhenius plot (that is, $\ln k$ versus $1/T$) yield a straight line?

(a) From 10°C to 35°C, the reaction rate increases with temperature as expected according to the Arrhenius equation. Above 35°C, the enzyme denatures, losing its catalytic ability, and the reaction rate slows.

(b) Simple Arrhenius behavior, resulting in a straight-line Arrhenius plot, occurs only for reactions whose rate is governed by a single rate constant. Thus, Equation 10.8, which contains the three rate constants, k_1, k_{-1}, and k_2, will not yield a straight line. At high substrate concentration, however, Equation 13.8 becomes $v_0 = k_2[E]_0[S]$, and a plot of $\ln k_{observed}$ vs. $1/T$ will give a straight line with slope related to the activation energy of the second (product forming) step.

10.32 Give an explanation for the Lineweaver–Burk plot for a certain enzyme-catalyzed reaction show below.

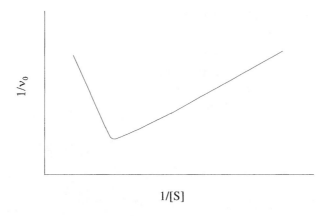

The plot shows that at high substrate concentration (low values of 1/[S]), the initial rate of the reaction decreases ($1/v_0$ increases). Thus, the substrate must act as an inhibitor to the enzyme.

10.34   In Lewis Carroll's tale "Through the Looking Glass," Alice wonders whether looking-glass milk on the other side of the mirror would be fit to drink. What do you think?

If Alice were to drink the milk, she would at the very least suffer the symptoms of lactose intolerance that affect people who have insufficient levels of lactase, the enzyme necessary to properly digest this milk sugar. This would occur because her lactase only acts on one of the two enantiomers of the compound, not the mirror image lactose present in looking glass milk. There are certainly other optically isomeric compounds present in milk that would also affect its fitness for Alice's consumption.

10.36   When fruits such as apples and pears are cut, the exposed areas begin to turn brown. This is the result of an oxidation reaction. Often the browning action can be prevented or slowed by adding a few drops of lemon juice. What is the chemical basis for this treatment?

The discoloration is due to an enzyme-catalyzed air oxidation of phenolic compounds in these fruits to form brown looking polymers. The citric acid in lemon juice deactivates the enzyme and prevents the oxidation.

10.38   Despite what you may have read in science fiction novels or seen in horror movies, it is extremely unlikely that insects can ever grow to human size. Why?

Insects have blood that contains no hemoglobin. Without a means of efficient transport for O_2 and CO_2, insects are limited to small sizes so that their metabolic processes can obtain enough O_2 via diffusion through their circulating system. Because of this limitation, they cannot grow to horror movie size.

10.40 Referring to Equation 10.12, sketch the Eadie–Hofstee plots for **(a)** a competitive inhibitor and **(b)** a noncompetitive inhibitor.

The plots are constructed using the following relations:

No inhibitor:

$$v_0 = \frac{V_{\text{max}}[S]}{K_M + [S]}$$

Competitive inhibitor:

$$v_0 = \frac{V_{\text{max}}[S]}{K_M\left(1 + \frac{[I]}{K_I}\right) + [S]}$$

Noncompetitive inhibitor:

$$v_0 = \frac{\frac{V_{\text{max}}}{\left(1 + \frac{[I]}{K_I}\right)}[S]}{K_M + [S]}$$

In the following curves, $[I] = 0$ denotes the absence of an inhibitor and $[I_1] < [I_2]$.

(a) Competitive inhibitor:

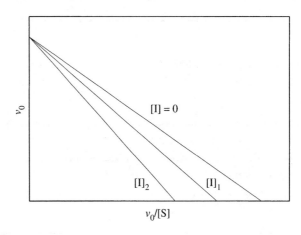

(b) Noncompetitive inhibitor

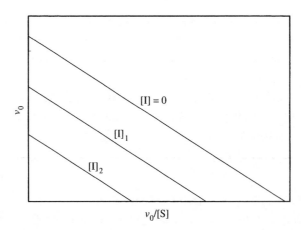

CHAPTER 11

Quantum Mechanics and Atomic Structure

PROBLEMS AND SOLUTIONS

11.2 The threshold frequency for dislodging an electron from a zinc metal surface is 8.54×10^{14} Hz. Calculate the minimum amount of energy required to remove an electron from the metal.

At the threshold frequency, the speed of electron, v, is 0. That is,

$$h\nu = \Phi + \frac{1}{2}m_e v^2 = \Phi$$

Therefore, the minimum amount of energy required to remove an electron from the metal is

$$\Phi = \left(6.626 \times 10^{-34}\,\text{J s}\right)\left(8.54 \times 10^{14}\,\text{Hz}\right) = 5.66 \times 10^{-19}\,\text{J}$$

11.4 Calculate the frequency and wavelength associated with the transition from the $n = 5$ to the $n = 3$ level in atomic hydrogen.

The wavenumber of the emitted radiation is

$$\tilde{\nu} = \left(109737\,\text{cm}^{-1}\right)\left|\left(\frac{1}{n_i^2} - \frac{1}{n_f^2}\right)\right| = \left(109737\,\text{cm}^{-1}\right)\left|\left(\frac{1}{5^2} - \frac{1}{3^2}\right)\right| = 7803.520\,\text{cm}^{-1}$$

Therefore, the wavelength and the frequency of this radiation are

$$\lambda = \frac{1}{\tilde{\nu}} = \frac{1}{7803.520\,\text{cm}^{-1}} = 1.28147 \times 10^{-4}\,\text{cm} = 1.28147 \times 10^3\,\text{nm}$$

$$\nu = c\tilde{\nu} = \left(3.00 \times 10^{10}\,\text{cm s}^{-1}\right)\left(7803.520\,\text{cm}^{-1}\right) = 2.34 \times 10^{14}\,\text{Hz}$$

11.6 A photoelectric experiment was performed by separately shining a laser at 450 nm (blue light) and a laser at 560 nm (yellow light) on a clean metal surface and measuring the number and kinetic energy of the ejected electrons. Which light would generate more electrons? Which light would eject electrons with greater kinetic energy? Assume that the same number of photons is delivered to the metal surface by each laser and that the frequencies of the laser lights exceed the threshold frequency.

Since each laser is delivering the same number of photons to the metal surface, the two experiments will each generate the same number of electrons. Since each blue light photon has greater energy than each yellow light photon, those electrons ejected by the blue light will have greater kinetic energy.

11.8 In a photoelectric experiment, a student uses a light source whose frequency is greater than that needed to eject electrons from a certain metal. After continuously shining the light on the same area of the metal for a long period of time, however, the student notices that the maximum kinetic energy of ejected electrons begins to decrease, even though the frequency of the light is held constant. How would you account for this behavior?

In an photoelectric experiment where the ejected electrons are not replaced (perhaps by making the photocathode part of an electric circuit), the metal surface will become positively charged due to the loss of the negatively charged electrons. Eventually, this positive charge is sufficient to cause a noticeable attraction between the surface and the ejected electrons which lowers their kinetic energy.

11.10 Suppose that the uncertainty in determining the position of an electron circling an atom in an orbit is 0.4 Å. What is the uncertainty in its velocity?

The uncertainty in momentum is

$$\Delta p \geq \frac{h}{4\pi\,\Delta x} = \frac{6.626 \times 10^{-34}\ \text{J s}}{4\pi\,\left(0.4 \times 10^{-10}\ \text{m}\right)} = 1.3 \times 10^{-24}\ \text{kg m s}^{-1}$$

Therefore, the uncertainty in the velocity of the electron is

$$\Delta v = \frac{\Delta p}{m} \geq \frac{1.3 \times 10^{-24}\ \text{kg m s}^{-1}}{9.109 \times 10^{-31}\ \text{kg}} = 1 \times 10^{6}\ \text{m s}^{-1}$$

The uncertainty principle, when applied to a microscopic object such as an electron, results in a significant uncertainty in velocity.

11.12 The diffraction phenomenon can be observed whenever the wavelength is comparable in magnitude to the size of the slit opening. To be "diffracted," how fast must a person weighing 84 kg move through a door 1 m wide?

The person would need a wavelength comparable to 1 m to be diffracted. The momentum of the person would be

$$p = \frac{h}{\lambda} = \frac{6.626 \times 10^{-34}\ \text{J s}}{1\ \text{m}} = 6.626 \times 10^{-34}\ \text{kg m s}^{-1}$$

The velocity of the person is therefore

$$v = \frac{p}{m} = \frac{6.626 \times 10^{-34}\ \text{kg m s}^{-1}}{84\ \text{kg}} = 7.9 \times 10^{-36}\ \text{m s}^{-1}$$

At this rate, it would take 1.3×10^{35} s or 4.1×10^{27} years to move 1 m!

11.14 The He$^+$ ion contains only one electron and is therefore a hydrogenlike ion. Calculate the wavelengths, in increasing order, of the first four transitions in the Balmer series of the He$^+$ ion. Compare these wavelengths with the same transitions in a H atom. Comment on the differences. (The Rydberg constant for He$^+$ is 8.72×10^{-18} J.)

Convert the Rydberg constant for He$^+$ into cm^{-1}:

$$R_H = \frac{8.72 \times 10^{-18} \text{ J}}{hc} = \frac{8.72 \times 10^{-18} \text{ J}}{\left(6.626 \times 10^{-34} \text{ J s}\right)\left(3.00 \times 10^8 \text{ m s}^{-1}\right)}$$

$$= 4.387 \times 10^7 \text{ m}^{-1} = 4.387 \times 10^5 \text{ cm}^{-1}$$

The wavenumbers and wavelengths for first four transitions (n_i = 3, 4, 5, 6) in the Balmer series can be calculated using

$$\tilde{\nu} = R_H \left| \left(\frac{1}{n_i^2} - \frac{1}{2^2} \right) \right|$$

$$\lambda = \frac{1}{\tilde{\nu}}$$

where R_H = 109737 cm^{-1} for H and 4.387×10^5 cm^{-1} for He$^+$. The wavelengths are

n_i	λ in nm for He$^+$	λ in nm for H
3	164	656
4	122	486
5	109	434
6	103	410

All the Balmer transitions for He$^+$ are in the ultraviolet region whereas the transitions for H are all in the visible region. The wavelength for a transition in He$^+$ is 1/4 that of the corresponding transition in H due to the factor of Z^2 in the R_H expression.

11.16 The retina of a human eye can detect light when radiant energy incident on it is at least 4.0×10^{-17} J. For light of 600-nm wavelength, how many photons does this correspond to?

The energy of a single 600 nm photon is

$$E = h\nu$$

$$= \frac{hc}{\lambda}$$

$$= \frac{\left(6.626 \times 10^{-34} \text{ J s}\right)\left(3.00 \times 10^8 \text{ m s}^{-1}\right)}{600 \times 10^{-9} \text{ m}}$$

$$= 3.313 \times 10^{-19} \text{ J}$$

The number of these photons required to provide 4.0×10^{-17} J so that the light can be detected by human eyes is

$$\frac{4.0 \times 10^{-17} \text{ J}}{3.313 \times 10^{-19} \text{ J}} = 1.2 \times 10^2$$

11.18   Ozone (O_3) in the stratosphere absorbs the harmful radiation from the sun by undergoing decomposition: $O_3 \rightarrow O + O_2$. **(a)** Referring to Appendix 2, calculate the $\Delta_r H^\circ$ value for this process. **(b)** Calculate the maximum wavelength of photons (in nm) that possess this energy to bring about the decomposition of ozone photochemically.

(a)

$$\Delta_r H^\circ = \Delta_f \overline{H}^\circ \left[O(g) \right] + \Delta_f \overline{H}^\circ \left[O_2(g) \right] - \Delta_f \overline{H}^\circ \left[O_3(g) \right]$$

$$= 249.4 \text{ kJ mol}^{-1} + 0 \text{ kJ mol}^{-1} - 142.7 \text{ kJ mol}^{-1}$$

$$= 106.7 \text{ kJ mol}^{-1}$$

(b) Assuming that the decomposition requires a single photon and proceeds with 100% efficiency, the decomposition of 1 mole of O_3 requires 1 mole of photons. The energy of one photon is

$$E = h\nu = \frac{hc}{\lambda}$$

so that the energy of 1 mole of photons is

$$E = \frac{hc}{\lambda} \left(6.022 \times 10^{23} \right)$$

The wavelength required so that the photons possess the necessary energy is

$$\lambda = \frac{hc}{E} \left(6.022 \times 10^{23} \right) = \frac{\left(6.626 \times 10^{-34} \text{ J s} \right) \left(3.00 \times 10^8 \text{ m s}^{-1} \right)}{106.7 \times 10^3 \text{ J}} \left(6.022 \times 10^{23} \right)$$

$$= 1.12 \times 10^{-6} \text{ m} = 1.12 \times 10^3 \text{ nm}$$

11.20   A student records an emission spectrum of hydrogen and notices that one spectral line in the Balmer series cannot be accounted for by the Bohr theory. Assuming that the gas sample is pure, suggest a species that might be responsible for this line.

The line is most likely due to the emission of H_2.

11.22   How many photons at 660 nm must be absorbed to melt 5.0×10^2 g of ice? On average, how many H_2O molecules does one photon convert from ice to water? (*Hint*: It takes 334 J to melt 1 g of ice at 0°C.)

The amount of energy that must be absorbed to melt 5.0×10^2 g of ice is

$$\left(5.0 \times 10^2 \text{ g} \right) \left(334 \text{ J g}^{-1} \right) = 1.67 \times 10^5 \text{ J}$$

The energy of 1 photon at 660 nm is

$$h\nu = \frac{hc}{\lambda} = \frac{\left(6.626 \times 10^{-34} \text{ J s} \right) \left(3.00 \times 10^8 \text{ m s}^{-1} \right)}{660 \times 10^{-9} \text{ m}} = 3.012 \times 10^{-19} \text{ J}$$

so that the number of photons required to melt the water is

$$\frac{1.67 \times 10^5 \text{ J}}{3.012 \times 10^{-19} \text{ J}} = 5.54 \times 10^{23} = 5.5 \times 10^{23}$$

The number of H_2O molecules converted in the 5.0×10^2 g sample from ice to water is

$$\frac{5.0 \times 10^2 \text{ g}}{18.0 \text{ g mol}^{-1}} \frac{6.022 \times 10^{23}}{1 \text{ mol}} = 1.67 \times 10^{25}$$

Since it took 5.54×10^{23} photons to melt the entire sample, the number of H_2O molecules converted from ice to water by 1 photon is

$$\frac{1.67 \times 10^{25}}{5.54 \times 10^{23}} = 30$$

11.24 According to Equation 11.22, the energy is inversely proportional to the square of the length of the box. How would you account for this dependence in terms of the Heisenberg uncertainty principle?

As the length of the box is decreased, the particle is located with less uncertainty. According to the uncertainty principle,

$$\Delta x \Delta p \geq \frac{h}{4\pi}$$

a decrease in Δx, here due to a shortening of the box, requires an increase in Δp and hence p itself. Consequently, the kinetic energy of the particle, $p^2/2m$ must also increase.

11.26 Derive Equation 11.22 using de Broglie's relation. (*Hint:* First you must express the wavelength of the particle in the nth level in terms of the length of the box.)

According to Figure 11.18(a), the wavelength of the particle is given by

$$\lambda = \frac{2L}{n}$$

where $n = 1, 2, 3...$ The wavelength is also related to the momentum of the particle:

$$\lambda = \frac{h}{p} = \frac{h}{mv}$$

Equating the two expressions for λ and solving for v gives

$$\frac{2L}{n} = \frac{h}{mv}$$

$$v = \frac{nh}{2mL}$$

The particle has only kinetic energy so that

$$E = \frac{1}{2}mv^2 = \frac{1}{2}m\frac{n^2h^2}{4m^2L^2} = \frac{n^2h^2}{8mL^2}$$

11.28 Based on the particle-in-a-one-dimensional-box model, suggest where along the box the $n = 1 \rightarrow n = 2$ electronic transition would most likely take place. Explain your choice.

For a transition to take place, both initial and final states must have non-zero probabilities of finding the electron at a given location. Consequently, the transition must take place at points where both ψ_1^2 and ψ_2^2 are non-zero. A likely place is shown in the figure.

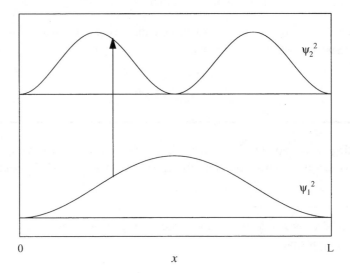

11.30 Obtain an expression for the most probable radius at which an electron will be found when it occupies the $1s$ orbital.

The $1s$ radial distribution function is

$$f = 4\pi r^2 R(r)^2 = 4\pi r^2 \left(\frac{2}{\sqrt{a_0^3}} e^{-r/a_0} \right)^2 = \frac{16\pi}{a_0^3} r^2 e^{-2r/a_0}$$

The radial distribution function reaches a maximum with $df/dr = 0$ at the most probable radius, r_{mp}. Differentiating f with respect to r,

$$\frac{df}{dr} = \frac{16\pi}{a_0^3} \left[2r e^{-2r/a_0} + r^2 \left(-\frac{2}{a_0} \right) e^{-2r/a_0} \right]$$

$r = r_{mp}$ when $\dfrac{df}{dr} = 0$.

$$\frac{16\pi}{a_0^3} \left[2r_{mp} e^{-2r_{mp}/a_0} + r_{mp}^2 \left(-\frac{2}{a_0} \right) e^{-2r_{mp}/a_0} \right] = 0$$

$$2 - \frac{2r_{mp}}{a_0} = 0$$

$$r_{mp} = a_0$$

The most probable radius for the $1s$ orbital is the Bohr radius.

11.32 Write the ground-state electron configurations of the following ions, which play important roles in biochemical processes in our bodies: **(a)** Na^+, **(b)** Mg^{2+}, **(c)** Cl^-, **(d)** K^+, **(e)** Ca^{2+}, **(f)** Fe^{2+}, **(g)** Cu^{2+}, **(h)** Zn^{2+}.

(a) Na^+: $1s^2 2s^2 2p^6 = [Ne]$

(b) Mg^{2+}: $1s^2 2s^2 2p^6 = [Ne]$

(c) Cl^-: $1s^2 2s^2 2p^6 3s^2 3p^6 = [Ar]$

(d) K^+: $1s^2 2s^2 2p^6 3s^2 3p^6 = [Ar]$

(e) Ca^{2+}: $1s^2 2s^2 2p^6 3s^2 3p^6 = [Ar]$

(f) Fe^{2+}: $1s^2 2s^2 2p^6 3s^2 3p^6 3d^6 = [Ar]3d^6$

(g) Cu^{2+}: $1s^2 2s^2 2p^6 3s^2 3p^6 3d^9 = [Ar]3d^9$

(h) Zn^{2+}: $1s^2 2s^2 2p^6 3s^2 3p^6 3d^{10} = [Ar]3d^{10}$

11.34 Ionization energy is the energy required to remove a ground state ($n = 1$) electron from an atom. It is usually expressed in units of $kJ\,mol^{-1}$. **(a)** Calculate the ionization energy for the hydrogen atom. **(b)** Repeat the calculation, assuming in this case that the electron is removed from the $n = 2$ state.

The ionization energy for the hydrogen atom ($Z = 1$) can be calculated using

$$\Delta E = \left(\frac{m_e Z^2 e^4}{8h^2 \epsilon_0^2} \right) \left(\frac{1}{n_i^2} - \frac{1}{n_f^2} \right)$$

$$= \left(2.1799 \times 10^{-18}\,J \right) \left(\frac{1}{n_i^2} - \frac{1}{n_f^2} \right)$$

with $n_f = \infty$. Therefore, the ionization energy, IE, for one hydrogen atom is

$$IE = \left(2.1799 \times 10^{-18}\,J \right) \left(\frac{1}{n_i^2} \right)$$

(a) If $n_i = 1$, for the hydrogen atom,

$$IE = \left(2.1799 \times 10^{-18}\,J \right) \left(\frac{1}{1} \right) = 2.1799 \times 10^{-18}\,J$$

For 1 mole of such hydrogen atoms,

$$IE = \left(2.1799 \times 10^{-18}\,J \right) \left(\frac{6.022 \times 10^{23}}{1\,mol} \right) = 1.313 \times 10^6\,J\,mol^{-1} = 1.313 \times 10^3\,kJ\,mol^{-1}$$

(b) If $n_i = 2$, for the hydrogen atom,

$$IE = \left(2.1799 \times 10^{-18}\,J \right) \left(\frac{1}{4} \right) = 5.44975 \times 10^{-19}\,J$$

For 1 mole of such hydrogen atoms,

$$\text{IE} = \left(5.44975 \times 10^{-19}\,\text{J}\right)\left(\frac{6.022 \times 10^{23}}{1\,\text{mol}}\right) = 3.282 \times 10^5\,\text{J mol}^{-1} = 3.282 \times 10^2\,\text{kJ mol}^{-1}$$

11.36   A technique called photoelectron spectroscopy is used to measure the ionization energy of atoms. A sample is irradiated with UV light, which causes electrons to be ejected from the valence shell. The kinetic energies of the ejected electrons are measured. Knowing the energy of the UV photon and the kinetic energy of the ejected electron, we can write

$$h\nu = \text{IE} + \frac{1}{2}m_\text{e}v^2$$

where ν is the frequency of the UV light, and m_e and v are the mass and velocity of the electron, respectively. In one experiment, the kinetic energy of the ejected electron from potassium is found to be 5.34×10^{-19} J using a UV source of wavelength 162 nm. Calculate the ionization energy of potassium. How can you be sure that this ionization energy corresponds to the electron in the valence shell (that is, the most loosely held electron)?

The ionization energy for one potassium atom is

$$\text{IE} = h\nu - \frac{1}{2}m_\text{e}v^2$$

$$= \frac{hc}{\lambda} - \text{KE}$$

$$= \frac{\left(6.626 \times 10^{-34}\,\text{J s}\right)\left(3.00 \times 10^8\,\text{m s}^{-1}\right)}{162 \times 10^{-9}\,\text{m}} - 5.34 \times 10^{-19}\,\text{J}$$

$$= 6.930 \times 10^{-19}\,\text{J}$$

The ionization energy for one mole of potassium atoms is

$$\text{IE} = \left(6.930 \times 10^{-19}\,\text{J}\right)\left(\frac{6.022 \times 10^{23}}{1\,\text{mol}}\right)$$

$$= 4.173 \times 10^5\,\text{J mol}^{-1} = 4.173 \times 10^2\,\text{kJ mol}^{-1}$$

To ensure that the ejected electron is the valence electron, UV light of the longest wavelength (lowest energy) should be used that can still eject electrons.

11.38   Experimentally, the electron affinity of an element can be determined by using a laser light to ionize the anion of the element in the gas phase:

$$\text{X}^-(g) + h\nu \rightarrow \text{X}(g) + \text{e}^-$$

Referring to Table 11.5, calculate the photon wavelength (in nanometers) corresponding to the electron affinity for chlorine. In what region of the electromagnetic spectrum does this wavelength fall?

The electron affinity for Cl is 349 kJ mol^{-1}. That is,

$$\text{Cl}(g) + \text{e}^- \rightarrow \text{Cl}^-(g) \qquad \Delta H = -349\,\text{kJ mol}^{-1}$$

The reverse reaction

$$Cl^-(g) + h\nu \rightarrow Cl(g) + e^-$$

occurs when the photon energy = 349 kJ mol^{-1}. The energy of one photon is

$$E = \left(349 \text{ kJ mol}^{-1}\right) \left(\frac{1 \text{ mol}}{6.022 \times 10^{23}}\right) = 5.795 \times 10^{-22} \text{ kJ} = 5.795 \times 10^{-19} \text{ J}$$

Since

$$E = h\nu = \frac{hc}{\lambda}$$

The wavelength corresponding to this energy is

$$\lambda = \frac{hc}{E}$$

$$= \frac{\left(6.626 \times 10^{-34} \text{ J s}\right)\left(3.00 \times 10^8 \text{ m s}^{-1}\right)}{5.795 \times 10^{-19} \text{ J mol}^{-1}}$$

$$= 3.43 \times 10^{-7} \text{ m} = 343 \text{ nm}$$

This wavelength is in the UV region.

11.40 Explain why the electron affinity of nitrogen is approximately zero, while the elements on either side, carbon and oxygen, have substantial positive electron affinities.

The electron affinity depends on the Z_{eff} for the *empty* orbital into which the additional electron is placed. In general, Z_{eff} increases across a row in the periodic table, so that electrons are held more tightly and increasing electron affinity. Thus, carbon and oxygen have electron affinities on the order of 100 kJ mol^{-1}. In the case of nitrogen, however, the additional electron must go into an orbital that is already half-occupied, and breaks up the half-filled subshell. Consequently, there is little tendency for the atom to accept another electron.

11.42 The first four ionization energies of an element are approximately 738 kJ mol^{-1}, 1450 kJ mol^{-1}, 7.7×10^3 kJ mol^{-1}, and 1.1×10^4 kJ mol^{-1}. To which periodic group does this element belong? Why?

The large jump between the second and third ionization energies indicates a change in the principal quantum number n. That is, if the first two electron removed have principal quantum number n, then the next two have principal quantum number $n - 1$. Thus, the element is in the second column of the periodic table, or Group 2A.

11.44 Photodissociation of water,

$$H_2O(g) + h\nu \rightarrow H_2(g) + \frac{1}{2} O_2(g)$$

has been suggested as a source of hydrogen. The $\Delta_r H^\circ$ value for the reaction, calculated from thermochemical data, is 285.8 kJ per mole of water decomposed. Calculate the maximum wavelength (in nm) that would provide the necessary energy. In principle, is it feasible to use sunlight as a source of energy for this process?

The minimum photon energy is 285.8 kJ mol^{-1}.

$$E = \left(285.8 \times 10^3 \text{ J mol}^{-1}\right) \left(\frac{1 \text{ mol}}{6.022 \times 10^{23}}\right) = 4.746 \times 10^{-19} \text{ J} = h\nu = \frac{hc}{\lambda}$$

$$\lambda = \frac{\left(6.626 \times 10^{-34} \text{ J s}\right)\left(3.00 \times 10^8 \text{ m s}^{-1}\right)}{4.746 \times 10^{-19} \text{ J}} = 4.19 \times 10^{-7} \text{ m} = 419 \text{ nm}$$

This wavelength is in the visible range of the electromagnetic spectrum. Since water is continuously being struck by visible radiation without decomposition, it seems unlikely that photodissociation of water by this method is possible.

11.46   Only a fraction of the electrical energy supplied to a tungsten light bulb is converted to visible light. The rest of the energy shows up as infrared radiation (that is, heat). A 75-W light bulb converts 15.0% of the energy supplied to it into visible light. Assuming a wavelength of 550 nm, how many photons are emitted by the light bulb per second? (1 W = 1 J s^{-1}.)

The energy of visible light emitted by the light bulb per second is

$$E_{\text{bulb}} = \left(75 \text{ J s}^{-1}\right)(1 \text{ s})(15\%) = 11.3 \text{ J}$$

The energy of a 550 nm photon is

$$E_{\text{photon}} = h\nu = \frac{hc}{\lambda} = \frac{\left(6.626 \times 10^{-34} \text{ J s}\right)\left(3.00 \times 10^8 \text{ m s}^{-1}\right)}{550 \times 10^{-9} \text{ m}} = 3.614 \times 10^{-19} \text{ J}$$

The number of photons emitted by the light bulb per second is

$$\frac{11.3 \text{ J}}{3.614 \times 10^{-19} \text{ J}} = 3.1 \times 10^{19}$$

11.48   The ionization energy of a certain element is 412 kJ mol^{-1}. When the atoms of this element are in the first excited state, however, the ionization energy is only 126 kJ mol^{-1}. Based on this information, calculate the wavelength of light emitted in a transition from the first excited state to the ground state.

The ionization energy of 412 kJ mol^{-1} represents the energy difference between the ground state and the dissociation limit whereas the ionization energy of 126 kJ mol^{-1} represents the energy difference between the first excited state and the dissociation limit. Therefore, the energy difference between the ground state and the excited state is

$$\Delta E = (412 - 126) \text{ kJ mol}^{-1} = 286 \text{ kJ mol}^{-1}$$

The energy of light emitted in a transition from the first excited state to the ground state is therefore 286 kJ mol^{-1}. The wavelength emitted is calculated as follows.

$$E = \left(286 \times 10^3 \text{ J mol}^{-1}\right) \left(\frac{1 \text{ mol}}{6.022 \times 10^{23}}\right) = 4.749 \times 10^{-19} \text{ J} = h\nu = \frac{hc}{\lambda}$$

$$\lambda = \frac{\left(6.626 \times 10^{-34} \text{ J s}\right) \left(3.00 \times 10^8 \text{ m s}^{-1}\right)}{4.749 \times 10^{-19} \text{ J}} = 4.19 \times 10^{-7} \text{ m} = 419 \text{ nm}$$

11.50 In 1996, physicists created an antiatom of hydrogen. In such an atom, which is the antimatter equivalent of an ordinary atom, the electrical charges of all the component particles are reversed. Thus, the nucleus of an antiatom is made of an antiproton, which has the same mass as a proton but bears a negative charge, while the electron is replaced by an antielectron (also called positron) with the same mass as an electron but bearing a positive charge. Would you expect the energy levels, emission spectra, and atomic orbitals of an antihydrogen atom to be different from those of a hydrogen atom? What would happen if an antiatom of hydrogen collided with a hydrogen atom?

The antiatom of hydrogen should show the same characteristics with regard to energy levels, emission spectra, and atomic orbitals as does ordinary hydrogen. Should an antiatom of hydrogen collide with an ordinary hydrogen atom, they would annihilate each other, and energy would be given off.

11.52 Use Equation 2.28 to calculate the de Broglie wavelength of a N_2 molecule at 300 K.

The rms speed of N_2 is

$$v_{\text{rms}} = \sqrt{\frac{3RT}{\mathcal{M}}} = \sqrt{\frac{3 \left(8.314 \text{ J K}^{-1} \text{mol}^{-1}\right) (300 \text{ K})}{28.02 \times 10^{-3} \text{ kg}}} = 516.8 \text{ m s}^{-1}$$

Using $v = v_{\text{rms}}$, the de Broglie wavelength of the molecule is

$$\lambda = \frac{h}{mv} = \frac{6.626 \times 10^{-34} \text{ J s}}{(28.02 \text{ amu}) \left(1.661 \times 10^{-27} \text{ kg amu}^{-1}\right) \left(516.8 \text{ m s}^{-1}\right)} = 2.75 \times 10^{-11} \text{ m}$$

11.54 The sun is surrounded by a white circle of gaseous material called the corona, which becomes visible during a total eclipse of the sun. The temperature of the corona is in the millions of degrees Celsius, high enough to break up molecules and remove some or all of the electrons from atoms. One way astronomers have been able to estimate the temperature of the corona is by studying the emission lines of ions of certain elements. For example, the emission spectrum of Fe^{14+} ions has been recorded and analyzed. Knowing that it takes 3.5×10^4 kJ mol^{-1} to convert Fe^{13+} to Fe^{14+}, estimate the temperature of the sun's corona. (*Hint*: The average kinetic energy of 1 mole of a gas is $\frac{3}{2}RT$.)

The energy required to create the Fe^{14+} ion from Fe^{13+} must come from the thermal energy of the corona. That is, collisions with other species in the plasma provide the necessary energy.

Thus, estimate the average kinetic energy in the plasma as being equal to the ionization energy of Fe^{13+}.

$$KE = IE$$

$$\frac{3}{2}RT = 3.5 \times 10^7 \text{ J mol}^{-1}$$

$$T = \frac{2\left(3.5 \times 10^7 \text{ J mol}^{-1}\right)}{3\left(8.314 \text{ J K}^{-1} \text{ mol}^{-1}\right)} = 2.8 \times 10^6 \text{ K}$$

11.56   The equation for calculating the energies of the electron in a hydrogen atom or a hydrogenlike ion is given in $E_n = -\left(2.18 \times 10^{-18} \text{ J}\right) Z^2 \left(1/n^2\right)$, where Z is the atomic number of the element. One way to modify this equation for many-electron atoms is to replace Z with $(Z - \sigma)$, where σ is a positive dimensionless quantity called the shielding constant. Consider the helium atom as an example. The physical significance of σ is that it represents the extent of shielding that the two $1s$ electrons exert on each other. Thus the quantity $(Z - \sigma)$ is appropriately called the "effective nuclear charge." Use the first ionization energy of helium in Table 11.4 to calculate the value of σ.

The ionization energy for the helium atom $(Z = 2)$ can be calculated by using

$$\Delta E = \left[\frac{m_e\,(Z - \sigma)^2\,e^4}{8h^2\epsilon_0^2}\right]\left(\frac{1}{n_i^2} - \frac{1}{n_f^2}\right)$$

$$= \left(2.18 \times 10^{-18} \text{ J}\right)(2 - \sigma)^2\left(\frac{1}{n_i^2} - \frac{1}{n_f^2}\right)$$

with $n_f = \infty$. The experimental value of the ionization energy for an electron in He with $n = 1$ is used to determine σ.

$$IE = \left(2.18 \times 10^{-18} \text{ J}\right)(2 - \sigma)^2\left(\frac{1}{1}\right) = 3.98 \times 10^{-18} \text{ J}$$

$$\sigma = 0.649$$

11.58   The figure shown below represents the emission spectrum of a hydrogenlike ion in the gas phase. All the lines result from the electronic transitions from the excited states to the $n = 2$ state. **(a)** What electronic transitions correspond to lines B and C? **(b)** If the wavelength of line C is 27.1 nm, calculate the wavelengths of lines A and B. **(c)** Calculate the energy needed to remove the electron from the ion in the $n = 4$ state. **(d)** What is the physical significance of the continuum?

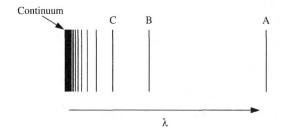

(a) Line A corresponds to the longest wavelength, lowest energy transition, which is the $n = 3 \rightarrow$ 2 transition. Lines B and C are the next two lines in the series to shorter wavelength, or higher energy. Therefore, line B corresponds to the $n = 4 \rightarrow 2$ transition, and line C corresponds to the $n = 5 \rightarrow 2$ transition.

(b) The wavelength of an electronic transition in a hydrogenlike ion is given by

$$\frac{1}{\lambda} = \tilde{\nu} = R_H Z^2 \left| \left(\frac{1}{n_i^2} - \frac{1}{n_f^2} \right) \right|$$

Consequently, the wavelengths of two electronic transitions in the ion are related by

$$\frac{\frac{1}{\lambda_1}}{\frac{1}{\lambda_2}} = \frac{\left| \left(\frac{1}{n_i^2} - \frac{1}{n_f^2} \right)_1 \right|}{\left| \left(\frac{1}{n_i^2} - \frac{1}{n_f^2} \right)_2 \right|}$$

or

$$\lambda_2 = \frac{\left| \left(\frac{1}{n_i^2} - \frac{1}{n_f^2} \right)_1 \right|}{\left| \left(\frac{1}{n_i^2} - \frac{1}{n_f^2} \right)_2 \right|} \lambda_1$$

Using the results of part **(a)**, for line A,

$$\lambda = \frac{\left| \left(\frac{1}{5^2} - \frac{1}{2^2} \right) \right|}{\left| \left(\frac{1}{3^2} - \frac{1}{2^2} \right) \right|} (27.1 \text{ nm}) = 41.0 \text{ nm}$$

and for line B

$$\lambda = \frac{\left| \left(\frac{1}{5^2} - \frac{1}{2^2} \right) \right|}{\left| \left(\frac{1}{4^2} - \frac{1}{2^2} \right) \right|} (27.1 \text{ nm}) = 30.4 \text{ nm}$$

(c) The wavelength of light necessary to remove the electron is given by the equation in part **(b)** with $n_i = 4$ and $n_f = \infty$

$$\lambda = \frac{\left| \left(\frac{1}{5^2} - \frac{1}{2^2} \right) \right|}{\left| \left(\frac{1}{4^2} - \frac{1}{\infty^2} \right) \right|} (27.1 \text{ nm}) = 91.06 \text{ nm}$$

This wavelength corresponds to a photon with an energy of

$$E = \frac{hc}{\lambda} = \frac{\left(6.626 \times 10^{-34} \text{ J s} \right) \left(3.00 \times 10^8 \text{ m s}^{-1} \right)}{91.06 \times 10^{-9} \text{ m}} = 2.18 \times 10^{-18} \text{ J}$$

(d) For higher energy levels in an atom or ion, the energy levels get closer together. Transitions from these levels with very high values of n to the $n = 2$ level will be very close in energy and hence will have similar wavelengths. The lines are so close together that they overlap, forming a continuum. The start of the continuum corresponds to the longest λ needed to ionize the ion.

Beyond that limit (to the left), the continuum corresponds to the movement of a free electron whose energy can vary continuously.

11.60   Use Equation 11.22 to account for the difference in the standard molar entropies of He and Ne.

Ne atoms are heavier than He atoms. Treating the atoms as particles in a box, Equation 11.22 requires that the energy levels for Ne be more closely spaced than those for He. (The same conclusion holds for 1-D box as for a 3-D box.) Thus, at a given temperature, there are more microstates associated with Ne than with He. Consequently, Ne has a larger standard molar entropy.

11.62   The nitrogen atom has one electron in each of the three $2p$ orbitals. Referring to Table 11.2, show that the total electron density is spherically symmetric, that is, it is independent of θ and ϕ. (*Hint*: Take the squares of the angular wave functions.)

The total electron density for the nitrogen atom is given by the sum of electron densities for each orbital. That is, the total electron density is the sum of the squares of the wave function for the occupied orbitals. Since the s orbitals are obviously spherical, only the sum over the three occupied $2p$ orbitals needs consideration. The radial portion of the wave function is common to all three orbitals.

$$\left|\psi_{2p_0}\right|^2 + \left|\psi_{2p_{+1}}\right|^2 + \left|\psi_{2p_{-1}}\right|^2$$

$$= R(r)^2 \left(\left| \sqrt{\frac{1}{2\pi}}\sqrt{\frac{3}{2}}\cos\theta \right|^2 + \left| \sqrt{\frac{1}{2\pi}}\sqrt{\frac{3}{4}}\sin\theta e^{+i\phi} \right|^2 + \left| \sqrt{\frac{1}{2\pi}}\sqrt{\frac{3}{4}}\sin\theta e^{-i\phi} \right|^2 \right)$$

$$= R(r)^2 \frac{3}{4\pi} \left(\cos^2\theta + \frac{1}{2}\sin^2\theta + \frac{1}{2}\sin^2\theta \right)$$

$$= R(r)^2 \frac{3}{4\pi} \left(\cos^2\theta + \sin^2\theta \right)$$

$$= R(r)^2 \frac{3}{4\pi}$$

The sum is seen to be independent of the angles θ and ϕ. In arriving at this result, we made use of the identities $\left|e^{\pm i\phi}\right|^2 = e^{\pm i\phi}e^{\mp i\phi} = e^0 = 1$, and $\sin^2\theta + \cos^2\theta = 1$.

11.64   Referring to Table 11.2, calculate the values of ρ for the $2s$ and $3s$ orbitals at which a node exists.

We can locate the nodes by setting the radial wavefunctions to 0 (see Problem 11.31). For the $2s$ orbital,

$$R(r) = 0 = \frac{1}{\sqrt{2a_0^3}} \left(1 - \frac{\rho}{2}\right) e^{-\rho/2}$$

$$1 - \frac{\rho}{2} = 0$$

$$\rho = 2$$

Since $\rho = \frac{r}{a_0}$, $\rho = 2$ is equivalent to $r = 2a_0$.

For the $3s$ orbital,

$$R(r) = 0 = \frac{2}{\sqrt{27a_0^3}} \left(1 - \frac{2}{3}\rho + \frac{2}{27}\rho^2\right) e^{-\rho/3}$$

$$1 - \frac{2}{3}\rho + \frac{2}{27}\rho^2$$

$$2\rho^2 - 18\rho + 27 = 0$$

$$\rho = 1.9 \text{ or } 7.1$$

These values correspond to $r = 1.9a_0$ and $7.1a_0$.

The results are consistent with the fact that the $2s$ orbital has one radial node while the $3s$ orbital has two radial nodes.

CHAPTER 12

The Chemical Bond

PROBLEMS AND SOLUTIONS

12.2 The chlorine nitrate molecule ($ClONO_2$) is believed to be involved in the destruction of ozone in the Antarctic stratosphere. Draw a plausible Lewis structure for this molecule.

Two equivalent resonance structures are possible for this molecule.

12.4 Carbon monoxide has a rather small dipole moment ($\mu = 0.12$ D) even though the electronegativity difference between C and O is rather large ($X_C = 2.5$ and $X_O = 3.5$). How would you explain this fact in terms of resonance structures?

The resonance form that makes the major contribution to the electronic structure of CO is $^-:C \equiv O:^+$. The lone pair and negative formal charge on the carbon tends to counteract the imbalance in shared electron pairs due to the greater electronegativity of oxygen. This results in a small dipole moment. (In fact, experiment shows that the carbon end of the molecule is the more negative end.)

12.6 Comment on the appropriateness of using the following resonance structure for O_2 intended to explain its paramagnetism.

This resonance structure indicates a single oxygen-to-oxygen bond in contradiction to the double bond that the molecule possesses. Furthermore, the oxygen atoms do not obey the octet rule.

12.8 Disulfide bonds play an important role in the three-dimensional structure of protein molecules. Discuss the nature of the –S–S– bond.

The electron configuration of the sulfur atom is $[Ne]3s^2 3p^4$, suggesting that each S atom is sp^3 hybridized to allow for each atom to form 2 covalent bonds and to have 2 lone pairs on each S atom as shown below.

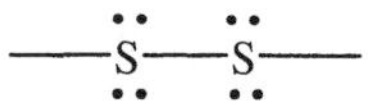

$$\overset{\cdot\cdot}{\underset{\cdot\cdot}{S}}-\overset{\cdot\cdot}{\underset{\cdot\cdot}{S}}$$

12.10 The unstable molecule carbene or methylene (CH_2) has been isolated and studied spectro-scopically. Suggest two types of bonding that might be present in this molecule. How would you determine which type of bond is present in CH_2?

One possibility would be for the C atom not to be hybridized. In this case there would be no unpaired electrons and the molecule would be diamagnetic.

A second possibility would be for the C atom to be sp^3 hybridized, with two of the hybrid orbitals used for bonding to the H's and the remaining two each containing a single, unpaired electron. These two unpaired electrons would cause the molecule to be paramagnetic.

This suggests that a measurement of the magnetic properties of the molecule would distinguish between the two cases. In fact, CH_2 in its ground state is paramagnetic and (nearly) linear. This would imply bonding to the H's using sp hybrids with a single, unpaired electron in each of the two unhybridized p orbitals.

12.12 Describe the bonding scheme in the following species in terms of molecular orbital theory: H_2^+, H_2, He_2^+, and He_2. List the species in order of decreasing stability.

The molecular orbital electron configuration for the four species is shown below. H_2 has a bond order of 1, both H_2^+ and He_2^+ have bond order of 1/2, and He_2 has a bond order of 0. Thus, the stability decreases as follows, $H_2 > H_2^+$, $He_2^+ > He_2$, with H_2^+ slightly more stable than He_2^+.

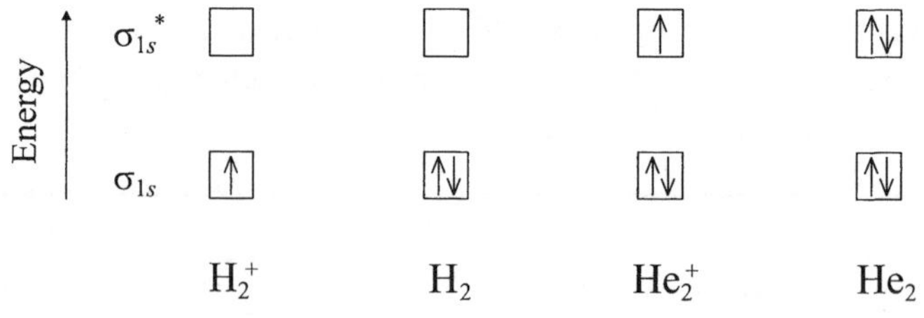

12.14 Which of the following species has the longest bond: CN^+, CN, or CN^-?

The electron configuration of CN is $KK(\sigma_{2s})^2 (\sigma_{2s}^*)^2 (\pi_x)^2 (\pi_y)^2 (\sigma_{2p})^1$. This results in a bond order of 2.5. In forming CN^+, an electron is removed from the bonding σ_{2p} orbital, which reduces the bond order to 2.0. When forming CN^-, the electron is added to the bonding σ_{2p} orbital, increasing the bond order to 3.0. Thus, CN^+ has the lowest bond order and the longest bond.

12.16 Which of the following two molecules has a greater degree of π-electron delocalization: naphthalene or biphenyl?

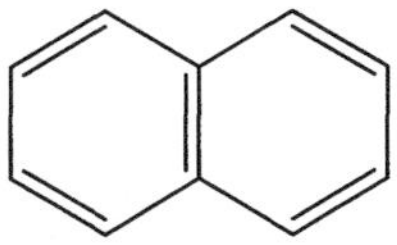

naphthalene 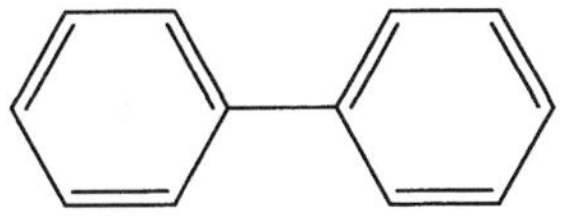   biphenyl

---

The rotation about the C–C single bond in biphenyl, shown below, partially destroys the electron delocalization between the two benzene rings, so that naphthalene has the greater degree of π-electron delocalization.

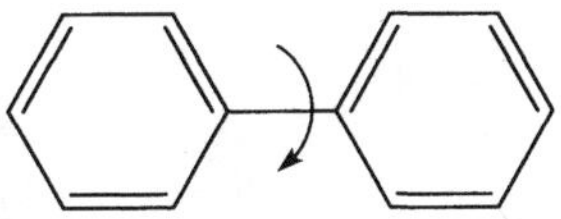

12.18 Compare the MO theory description for the H_2 molecule, where the wave function is given by

$$\psi = \left[\psi_A\,(1) + \psi_B\,(1)\right]\left[\psi_A\,(2) + \psi_B\,(2)\right]$$

with the VB theory treatment given by Equation 12.4. Under what condition do they become identical?

In Equation 12.4, the valence bond theory treatment gives the wave function

$$\psi_{VB} = \psi_A\,(1)\,\psi_B\,(2) + \psi_A\,(2)\,\psi_B\,(1) + \lambda\left[\psi_A\,(1)\,\psi_A\,(2) + \psi_B\,(1)\,\psi_B\,(2)\right]$$

The molecular orbital theory wave function as given above can be expanded to give

$$\psi_{MO} = \psi_A\,(1)\,\psi_B\,(2) + \psi_A\,(2)\,\psi_B\,(1) + \psi_A\,(1)\,\psi_A\,(2) + \psi_B\,(1)\,\psi_B\,(2)$$

The two wave functions are the same when $\lambda = 1$. The MO approach overemphasizes the ionic character, reckoning it equally as important as covalent character. In its first approximation, with $\lambda = 0$, the VB approach underestimates ionic character.

12.20 Describe the bonding in the nitrate ion, NO_3^-, in terms of delocalized molecular orbitals.

The ion contains 24 valence electrons. Of these, six are involved in three sigma bonds, indicated below, between the nitrogen and the oxygen atoms. The hybridization of the nitrogen atom is sp^2. There are four valence electrons on each oxygen atom, for a total of 12, which are non-bonding. The remaining six electrons are in delocalized π molecular orbitals which result from the overlap of the p_z orbital of the nitrogen with the p_z orbitals on the three oxygen atoms as shown below. The molecular orbitals are similar to those of the carbonate ion.

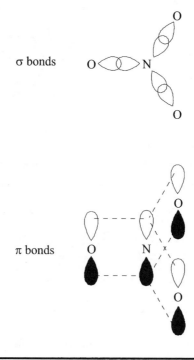

σ bonds

π bonds

12.22 Draw energy-level diagrams to show the low- and high-spin octahedral complexes of the transition-metal ions that have the electron configurations d^4, d^5, d^6, and d^7.

	Low spin		**High spin**

d^4

d^5

d^6

d^7

12.24   Predict the number of unpaired electrons in the following complex ions: **(a)** $Cr(CN)_6^{4-}$ and **(b)** $Cr(H_2O)_6^{2+}$.

(a) The electron configuration of Cr^{2+} is $[Ar]3d^4$. CN^- is a strong-field ligand, and the four electrons will occupy the three lower orbitals as shown. There will be two unpaired electrons.

[Cr(CN)₆]⁴⁻

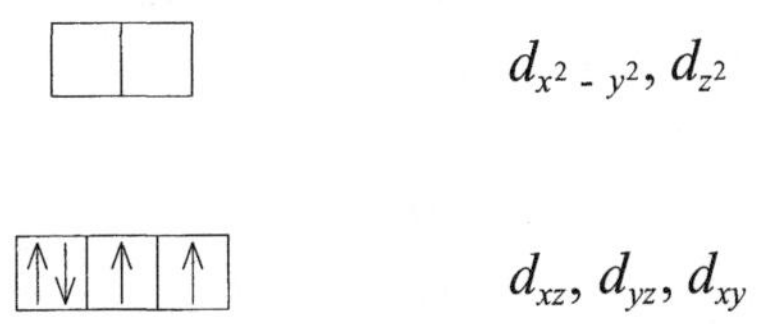

$d_{x^2-y^2}, d_{z^2}$

d_{xz}, d_{yz}, d_{xy}

(b) Since H_2O is a weak-field ligand, the four $3d$ electrons of Cr^{2+} will be arranged to maximize the number of unpaired electrons as shown below. There will be four unpaired electrons.

[Cr(H₂O)₆]²⁺

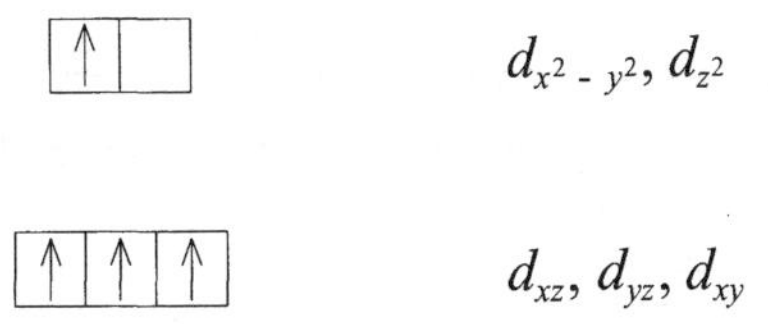

$d_{x^2-y^2}, d_{z^2}$

d_{xz}, d_{yz}, d_{xy}

12.26   The absorption maximum for the complex ion $Co(NH_3)_6^{3+}$ occurs at 470 nm. **(a)** Predict the color of the complex, and **(b)** calculate the crystal field splitting in $kJ\,mol^{-1}$.

(a) The complex ion is absorbing 470 nm light, which is blue to blue-green. Referring to Table 12.5, this implies that the complex is between yellow and red, or orange.

(b) The energy of a 470 nm photon is found using Planck's relationship and converted to a molar basis.

$$\Delta E = \frac{hc}{\lambda}$$
$$= \frac{\left(6.626 \times 10^{-34}\ J\,s\right)\left(3.00 \times 10^8\ m\,s^{-1}\right)}{470 \times 10^{-9}\ m}$$
$$= 4.229 \times 10^{-19}\ J$$

This is the crystal field splitting per molecule. Multiplying by Avogadro's constant gives the crystal field splitting for a mole of molecules.

$$\left(4.229 \times 10^{-19}\ J\right)\left(6.022 \times 10^{23}\ mol^{-1}\right)\left(\frac{1\ kJ}{1000\ J}\right) = 255\ kJ\,mol^{-1}$$

12.28   Hydrated Mn^{2+} ions are practically colorless even though they possess five $3d$ electrons. Explain. (*Hint*: Electronic transitions in which there is a change in the number of unpaired electrons do not occur readily.)

In a high-spin complex with five d electrons, all the electrons are unpaired and each of the five $3d$ orbitals are singly occupied as shown in the diagram below. Any excitation of a $3d$ electron in such a complex would require that an orbital become doubly occupied and that, to satisfy the Pauli exclusion principle, an electron would need to change its spin. Such transitions that involve a change in spin state are forbidden and do not occur to any appreciable extent. Thus, hydrated Mn^{2+} ions do not absorb light in the visible region of the spectrum and, as a result, appear faintly colored (pink).

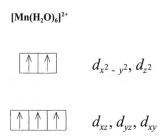

$[Mn(H_2O)_6]^{2+}$

12.30   Although both carbon and silicon are in Group 4A, very few Si=Si bonds are known. Account for the instability of silicon-to-silicon double bonds in general. (*Hint*: Compare the covalent radii of C and Si.)

The larger size of the Si atoms prevents effective sideways overlap of the $3p$ orbitals to form π bonds like those formed from the $2p$ orbitals in C atoms.

12.32   Chemical analysis shows that hemoglobin is 0.34% Fe by mass. What is the minimum possible molar mass of hemoglobin? The actual molar mass of hemoglobin is about 65,000 g. How do you account for the discrepancy between your minimum value and the actual value?

The mass percent indicates that a 100.00 g sample of hemoglobin contains 0.34 g Fe, or in moles,

$$(0.34\ \text{g}) \left(\frac{1\ \text{mol}}{55.85\ \text{g}} \right) = 6.09 \times 10^{-3}\ \text{mol}$$

There is a molar relationship between the moles of Fe and the moles of hemoglobin. The greatest possible number of moles of hemoglobin, implying the minimum possible molar mass, would be if there were 1 Fe atom per hemoglobin molecule. This would result in 100.00 g of hemoglobin likewise being 6.09×10^{-3} mol, or

$$\frac{100.00\ \text{g}}{6.09 \times 10^{-3}\ \text{mol}} = 1.6 \times 10^4\ \text{g mol}^{-1}$$

This is only a fraction of the actual molar mass of 65,000 g mol^{-1}, since there are more than 1 Fe atom per hemoglobin molecule. Indeed the molar mass ratio provides the Fe to hemoglobin ratio.

$$6.5 \times 10^4 \text{ g hemoglobin mol}^{-1} \left(\frac{1 \text{ mol Fe}}{1.6 \times 10^4 \text{ g hemoglobin}} \right) = 4.1 \text{ mol Fe (mol hemoglobin)}^{-1}$$

$$\approx 4 \text{ mol Fe (mol hemoglobin)}^{-1}$$

12.34 Co binds better to the heme group than Fe, and Co^{2+} has less of a tendency to be oxidized to Co^{3+} than Fe^{2+} does to Fe^{3+}. Why is Fe the metal in hemoglobin and myoglobin rather than Co?

Fe has a greater natural abundance in Earth's crust than does Co.

12.36 The dipole moment of *cis*-dichloroethylene is 1.81 D at 25°C. On heating, its dipole moment begins to decrease. Give a reasonable explanation for this observation.

Upon heating *cis-trans* isomerization takes place, and the *trans* isomer has no dipole moment.

12.38 The H_3^+ ion is the simplest polyatomic molecule. It has equilateral geometry. **(a)** Draw three resonance structures to represent this species. **(b)** Use MO theory to describe the bonding molecular orbital for this ion. Write the wave function for the lowest-energy molecular orbital. Is it a σ or π delocalized molecular orbital? **(c)** Given that $\Delta_r H = -849 \text{ kJ mol}^{-1}$ for the reaction $2H + H^+ \rightarrow H_3^+$ and that $\Delta_r H = 436.4 \text{ kJ mol}^{-1}$ for $H_2 \rightarrow 2H$, calculate the value of $\Delta_r H$ for the reaction $H^+ + H_2 \rightarrow H_3^+$. Comment on the magnitude of $\Delta_r H$.

(a)

(b) In the MO theory, a molecular orbital is formed from the overlap of the three $1s$ orbitals, one from each H, or $\psi = \frac{1}{\sqrt{3}} \left(1s_a + 1s_b + 1s_c \right)$. Although neither the σ nor π label is strictly appropriate for this molecule, since the orbital is formed from the overlap of s orbitals it could be termed a σ MO. This is an example of a three-center, two electron bond.

(c) This is an application of Hess's Law.

$$2H + H^+ \longrightarrow H_3^+ \qquad \Delta_r H = -849 \text{ kJ mol}^{-1}$$
$$H_2 \longrightarrow 2H \qquad \Delta_r H = 436.4 \text{ kJ mol}^{-1}$$

The two reactions sum to give

$$H^+ + H_2 \longrightarrow H_3^+ \qquad \Delta_r H = -413 \text{ kJ mol}^{-1}$$

The energy released in forming H_3^+ from H^+ and H_2 is almost as large as that released in the formation of H_2 from 2 H atoms. This result demonstrates the role of electron delocalization (σ electrons in this case) in producing stability in the H_3^+ ion.

12.40 Does the molecule HBrC=C=CHBr have a dipole moment?

The two CBrH groups are in different (perpendicular) planes (see figure), and the two bond moments will not cancel. Thus, HBrC=C=CHBr is a polar molecule.

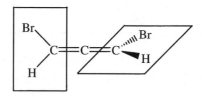

12.42 Oxalic acid, $H_2C_2O_4$, is sometimes used to remove rust stains from sinks and bathtubs. Explain the chemistry underlying this cleaning action.

The oxalate ion, from the oxalic acid, forms a water-soluble complex with the Fe^{3+} ion according to

$$Fe_2O_3(s) + 6H_2C_2O_4(aq) \longrightarrow 2Fe(C_2O_4)_3^{3-}(aq) + 3H_2O(l) + 6H^+(aq)$$

Note that bleach-based cleansers, which oxidize stains, would have little effect in removing a rust stain.

12.44 Draw qualitative diagrams for the crystal field splitting in **(a)** a linear complex ion ML_2, **(b)** a trigonal-planar complex ion ML_3, and **(c)** a trigonal-bipyramidal complex ion ML_5.

(a)

$$\text{-------- L—M—L --------} \rightarrow z$$

$\square$ d_{z^2}

$\square\square$ d_{xz}, d_{yz}

$\square\square$ $d_{x^2-y^2}, d_{xy}$

(b)

$$\begin{array}{c} L \\ \backslash \\ M—L \ \text{-----} \rightarrow x \\ / \\ L \end{array} \quad y$$

$\square\square$ $d_{x^2-y^2}, d_{xy}$

$\square$ d_{z^2}

$\square\square$ d_{xz}, d_{yz}

(c)

$$\begin{array}{c} L \\ | \\ L \cdots M—L \quad z \\ L \nearrow | \\ L \end{array}$$

$\square$ d_{z^2}

$\square\square$ $d_{x^2-y^2}, d_{xy}$

$\square\square$ d_{xz}, d_{yz}

12.46 The geometries discussed in this chapter all lend themselves to fairly straightforward elucidation of bond angles. The exception is the tetrahedron, because its bond angles are hard to visualize. Consider the CCl_4 molecule, which has tetrahedral geometry and is nonpolar. By equating the bond moment of a particular C–Cl bond to the resultant bond moments of the other three C–Cl bonds in opposite directions, show that the bond angles are all equal to 109.5°.

Referring to the figure below, if the bond moment of the upward-pointing C–Cl bond is represented by ρ, then for the molecule to be nonpolar, the downward-pointing components of the three other C–Cl bonds, each of which is $\rho \cos\theta$, must sum to give the same value. That is,

$$3\rho \cos\theta = \rho$$

$$\cos\theta = \frac{1}{3}$$

$$\theta = 70.5°$$

θ is the supplement of the tetrahedral angle, which is thus $180° - 70.5° = 109.5°$.

12.48 Cu^{2+} ions coordinated with S atoms tend to form tetrahedral complexes whereas those coordinated with N atoms tend to form octahedral complexes. Explain.

N is a smaller atom and is relatively high in the spectrochemical series whereas S is larger and relatively low in the spectrochemical series. There is less steric repulsion between ligands in a tetrahedral complex, and this geometry is favored by ligands containing S donor atoms. Octahedral complexes have a larger crystal field splitting (Δ) and hence a larger crystal field stabilization energy. Strong field ligands, such as those containing N donor atoms, favor this geometry because of the increased stabilization.

12.50 Which has a lower first ionization energy, O or O_2? Explain.

O_2 will have the lower first ionization energy because the electron is removed from an antibonding π^* orbital that is raised in energy in the molecule relative to the atom. Indeed, the experimental data are IE = 1314 kJ mol^{-1} for O, and IE = 1164.5 kJ mol^{-1} for O_2.

12.52 How many geometric isomers can the square-planar platinum complex Pt(abcd) have? Each letter represents a monodentate ligand.

There are 3 geometric isomers for the square planar platinum complex:

CHAPTER 13

Intermolecular Forces

PROBLEMS AND SOLUTIONS

13.2 Arrange the following species in order of decreasing melting points: Ne, KF, C_2H_6, MgO, H_2S.

$MgO > KF > H_2S > C_2H_6 > Ne$

13.4 If you lived in Alaska, which of the following natural gases would you keep in an outdoor storage tank in winter: methane (CH_4), propane (C_3H_8), or butane (C_4H_{10})? Explain.

CH_4 has the weakest intermolecular forces and, as a result, the lowest boiling point, making it the best choice for a cold climate.

13.6 The boiling points of the three different structural isomers of pentane (C_5H_{12}) are 9.5°C, 27.9°C, and 36.1°C. Draw their structures, and arrange them in order of decreasing boiling points. Justify your arrangement.

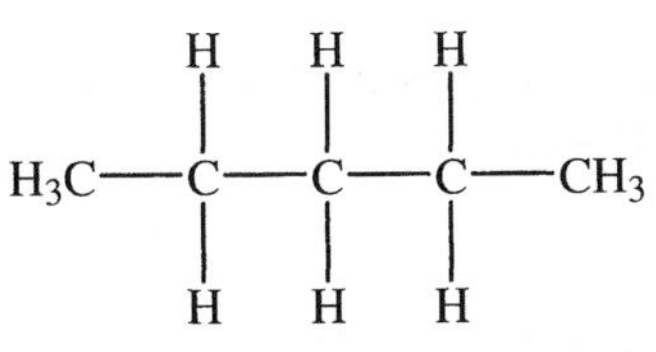

| n-pentane | 2-methylbutane | 2,2-dimethylpropane |
| 36.1 °C | 27.9 °C | 9.5 °C |

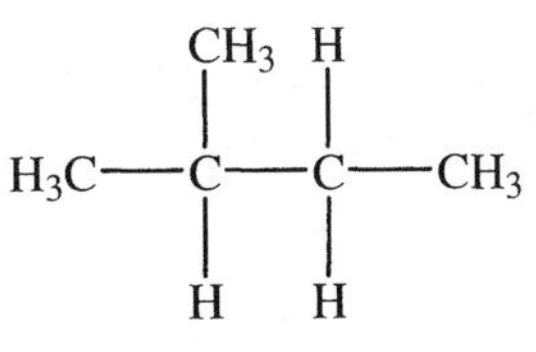

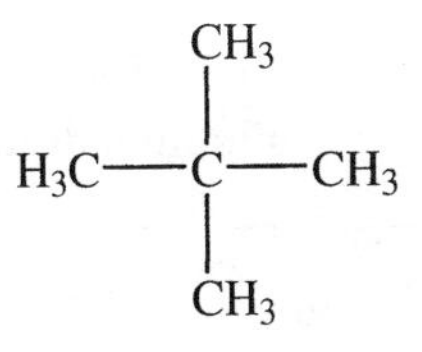

The boiling points depend on the ease of packing the molecules together. The n-pentane packs together most easily, and it has the highest boiling point. The packing is least favorable for 2,2-dimethylpropane, which has the lowest boiling point.

13.8 Coulombic forces are usually referred to as long-range forces (they depend on $1/r^2$), whereas van der Waals forces are called short-range forces (they depend on $1/r^7$). **(a)** Assuming that the forces (F) depend only on distances, plot F as a function of r at $r = 1$ Å, 2 Å, 3 Å, 4 Å, and 5 Å.

(b) Based on your results, explain the fact that although a 0.2 M nonelectrolyte solution usually behaves ideally, nonideal behavior is quite noticeable in a 0.02 M electrolyte solution.

(a) A plot with graphs of $1/r^2$ vs r and $1/r^7$ vs r is presented below. The forces will be proportional to these functions.

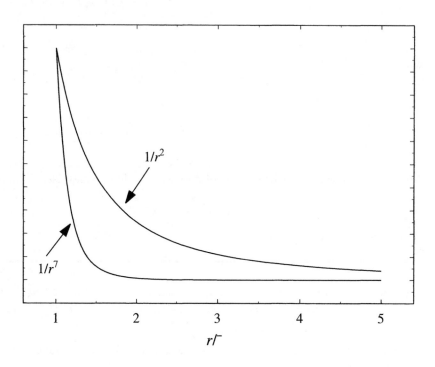

(b) In a nonelectrolyte solution, the attractive forces have a $1/r^7$ dependence, and as the graph shows, they fall off very rapidly with distance. In an electrolyte solution, the ionic (Coulombic) forces have a $1/r^2$ dependence that extends to large distances. These "long-range" forces are responsible for nonideal behavior, even at low concentrations.

13.10   Differentiate Equation 13.21 with respect to r to obtain an expression for σ and ϵ. Express the equilibrium distance, r_e, in terms of σ, and show that $V = -\epsilon$.

Starting with Equation 13.21,

$$V = 4\epsilon \left[\left(\frac{\sigma}{r} \right)^{12} - \left(\frac{\sigma}{r} \right)^{6} \right]$$

and differentiating gives

$$\frac{dV}{dr} = 4\epsilon \left[-\frac{12\sigma^{12}}{r^{13}} + \frac{6\sigma^6}{r^7} \right]$$

The minimum of the potential energy occurs when $r = r_e$ and $\dfrac{dV}{dr} = 0$.

$$4\epsilon\left[-\frac{12\sigma^{12}}{r_e^{13}}+\frac{6\sigma^6}{r_e^7}\right]=0$$

$$-\frac{12\sigma^{12}}{r_e^{13}}+\frac{6\sigma^6}{r_e^7}=0$$

$$-\frac{2\sigma^6}{r_e^6}+1=0$$

$$r_e=2^{1/6}\sigma$$

To calculate the potential energy at the equilibrium distance, substitute the expression for r_e into that for the potential energy.

$$V=4\epsilon\left[\left(\frac{\sigma}{2^{1/6}\sigma}\right)^{12}-\left(\frac{\sigma}{2^{1/6}\sigma}\right)^6\right]$$

$$=4\epsilon\left[\frac{1}{4}-\frac{1}{2}\right]$$

$$=-\epsilon$$

13.12 **(a)** From the data in Table 13.2, determine the van der Waals radius for argon. **(b)** Use this radius to determine the fraction of the volume occupied by 1 mole of argon at 25°C and 1 atm.

(a) Since σ gives the distance of closest approach of two argon atoms, the van der Waals radius for argon is

$$r=\frac{\sigma}{2}=\frac{3.40\text{ Å}}{2}=1.70\text{ Å}$$

(b) The volume of 1 mole of Ar atoms is

$$\frac{4}{3}\pi r^3\left(\frac{6.022\times10^{23}}{1\text{ mol}}\right)=\frac{4}{3}\pi\left(1.70\times10^{-10}\text{ m}\right)^3\left(\frac{6.022\times10^{23}}{1\text{ mol}}\right)$$

$$=1.239\times10^{-5}\text{ m}^3\text{ mol}^{-1}$$

$$=1.239\times10^{-2}\text{ L mol}^{-1}$$

The volume occupied by one mole of argon gas is

$$\frac{V}{n}=\frac{RT}{P}=\frac{\left(0.08206\text{ L atm K}^{-1}\text{ mol}^{-1}\right)(298.2\text{ K})}{1\text{ atm}}=24.47\text{ L mol}^{-1}$$

The fraction of this volume occupied by the one mole of argon atoms is

$$\frac{1.239\times10^{-2}\text{ L mol}^{-1}}{24.47\text{ L mol}^{-1}}=5.1\times10^{-4}$$

13.14 If water were a linear molecule, **(a)** would it still be polar and **(b)** would the water molecules still be able to form hydrogen bonds with one another?

(a) A "linear" water molecule would not be polar.

(b) Such a molecule could still form hydrogen bonds, although it would assume two-dimensional hydrogen bond structures.

13.16 Explain why ammonia is soluble in water but nitrogen trichloride is not.

Ammonia, NH_3, can form hydrogen bonds with water, but NCl_3 cannot.

13.18 Which of the following molecules has a higher melting point? Explain your answer.

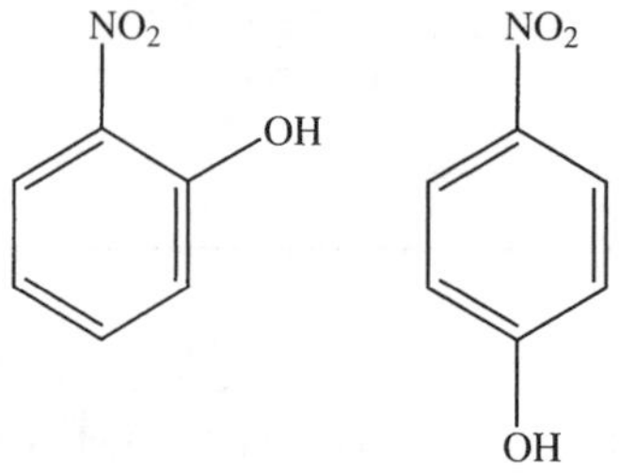

The *para* isomer can form intermolecular hydrogen bonds, while the *ortho* isomer can form only intramolecular hydrogen bonds as shown below. Thus, the *para* form with stronger intermolecular forces will have the higher melting point.

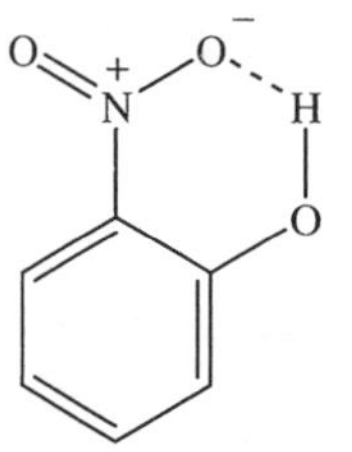

13.20 Assume the energy of hydrogen bonds per base pair to be 10 kJ mol^{-1}. Given two complementary strands of DNA containing 100 base pairs each, calculate the ratio of two separate strands to hydrogen-bonded double helix in solution at 300 K.

For one pair of bases, the ratio of the two separate strands to hydrogen-bonded double helix is

$$\exp\left(-\frac{\Delta E}{RT}\right) = \exp\left[-\frac{10 \times 10^3 \text{ J mol}^{-1}}{(8.314 \text{ J K}^{-1}\text{mol}^{-1})(300 \text{ K})}\right] = 1.8 \times 10^{-2}$$

For 100 base pairs, the ratio of the two separate strands to hydrogen-bonded double helix is

$$= \exp\left[-\frac{(100)\left(10 \times 10^3 \text{ J mol}^{-1}\right)}{(8.314 \text{ J K}^{-1}\text{mol}^{-1})(300 \text{ K})}\right] = 7.6 \times 10^{-175} \approx 0$$

13.22 List all the intra- and intermolecular forces that could exist between hemoglobin molecules in water.

All of the intermolecular interactions discussed in the chapter (dispersion, dipole–dipole, ionic) exist between hemoglobin molecules in water.

13.24 Which of the following properties indicates very strong intermolecular forces in a liquid? **(a)** A very low surface tension, **(b)** a very low critical temperature, **(c)** a very low boiling point, or **(d)** a very low vapor pressure.

Only **(d)** indicates very strong intermolecular forces in a liquid. The others indicate weak intermolecular forces.

13.26 Using values listed in Table 13.1 and a handbook of chemistry, plot the polarizabilities of the noble gases versus their boiling points. On the same graph, also plot their molar masses versus boiling points. Comment on the trends.

The necessary data are in the table. The polarizability of Rn is not accurately known; this noble gas is not included.

Noble Gas	Molar Mass/g·mol^{-1}	$\alpha/10^{-30} \cdot m^3$	b.p. / K
He	4.00	0.20	4.2
Ne	20.18	0.40	27.1
Ar	39.95	1.66	87.3
Kr	83.80	2.54	120.0
Xe	131.29	4.15	165.2

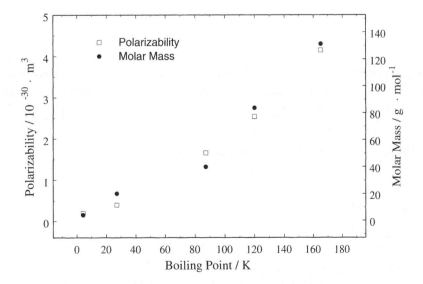

Both polarizability and molar mass seem to track the boiling point of the noble gas.

13.28   The HF_2^- ion exists as

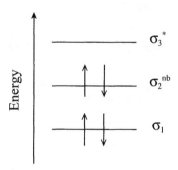

The fact that both HF bonds are the same length suggests that proton tunneling occurs. **(a)** Draw resonance structures for the ion. **(b)** Give a molecular orbital description (with an energy-level diagram) of hydrogen bonding in the ion.

(a)

(b) The $1s$ orbital on the H atom and a $2p$ orbital on each of the F atoms (the ones along the internuclear axis) combine to form three σ molecular orbitals: one bonding, one nonbonding, and one antibonding. There are four electrons to be accommodated in these molecular orbitals, and they are placed, paired, in the lowest two. Thus, there is a delocalized σ bond extending over the entire ion and a delocalized "lone pair" as a result of the nonbonding molecular orbital that has significant electron density at the fluorines, but a node at the hydrogen. (The other 12 valence electrons in the ion are in localized orbitals on the two fluorines.)

13.30   The potential energy of the helium dimer (He_2) is given by

$$V = \frac{B}{r^{13}} - \frac{C}{r^6}$$

where $B = 9.29 \times 10^4$ kJ $Å^{13}$ (mol dimer)$^{-1}$ and $C = 97.7$ kJ $Å^6$ (mol dimer)$^{-1}$. **(a)** Calculate the equilibrium distance between the He atoms. **(b)** Calculate the binding energy of the dimer. **(c)** Would you expect the dimer to be stable at room temperature (300 K)?

(a) The equilibrium distance, r_e, can be calculated by setting $\dfrac{dV}{dr} = 0$.

$$\frac{dV}{dr} = -\frac{13B}{r^{14}} + \frac{6C}{r^7}$$

$$-\frac{13B}{r_e^{14}} + \frac{6C}{r_e^7} = 0$$

$$r_e = \left(\frac{13B}{6C}\right)^{1/7} = \left\{ \frac{13\left[9.29 \times 10^4 \text{ kJ } Å^{13} \text{ (mol dimer)}^{-1}\right]}{6\left[97.7 \text{ kJ } Å^6 \text{ (mol dimer)}^{-1}\right]} \right\}^{1/7} = 2.975 \text{ Å} = 2.98 \text{ Å}$$

(b) The binding energy, $V(r_e)$, is

$$V(r_e) = \frac{B}{r_e^{13}} - \frac{C}{r_e^6}$$

$$= \frac{9.29 \times 10^4 \text{ kJ Å}^{13} \text{ (mol dimer)}^{-1}}{\left(2.975 \text{ Å}\right)^{13}} - \frac{97.7 \text{ kJ Å}^6 \text{ (mol dimer)}^{-1}}{\left(2.975 \text{ Å}\right)^6}$$

$$= -7.60 \times 10^{-2} \text{ kJ (mol dimer)}^{-1}$$

(c) The thermal energy at 300 K is

$$RT = \left(8.314 \text{ J K}^{-1} \text{ mol}^{-1}\right) (300 \text{ K}) = 2.49 \times 10^3 \text{ J mol}^{-1} = 2.49 \text{ kJ mol}^{-1}$$

which is much larger than 7.60×10^{-2} kJ (mol dimer)$^{-1}$. Thus the dimer would not be able to form at room temperature. This species has been observed and studied at low temperature.

CHAPTER 14

Spectroscopy

PROBLEMS AND SOLUTIONS

14.2 Convert 450 nm to wavenumber and frequency.

$$\lambda = \frac{c}{\nu} = \frac{1}{\tilde{\nu}}$$

$$\tilde{\nu} = \frac{1}{\lambda} = \frac{1}{450 \text{ nm}} \left(\frac{1 \text{ nm}}{1 \times 10^{-7} \text{ cm}} \right) = 2.2 \times 10^4 \text{ cm}^{-1}$$

$$\nu = \frac{c}{\lambda} = \frac{3.00 \times 10^8 \text{ m s}^{-1}}{450 \text{ nm}} \left(\frac{1 \text{ nm}}{1 \times 10^{-9} \text{ m}} \right) = 6.7 \times 10^{14} \text{ s}^{-1}$$

14.4 Convert the following absorbance to percent transmittance: **(a)** 0.0, **(b)** 0.12, and **(c)** 4.6.

$$T = \frac{I}{I_0} = 10^{\log \frac{I}{I_0}} = 10^{-A}$$

(a) $A = 0.0$, and $T = 10^{-0} = 1 = 100\%$.

(b) $A = 0.12$, and $T = 10^{-0.12} = 0.76 = 76\%$.

(c) $A = 4.6$, and $T = 10^{-4.6} = 2.5 \times 10^{-5} = 0.0025\%$.

14.6 The mean lifetime of an electronically excited molecule is 1.0×10^{-8} s. If the emission of the radiation occurs at 610 nm, what are the uncertainties in frequency ($\Delta\nu$) and wavelength ($\Delta\lambda$)?

The natural linewidth, or uncertainty in frequency, of a transition is related to the lifetime of the excited state through the uncertainty principle

$$\Delta\nu = \frac{1}{4\pi\,\Delta t}$$

$$= \frac{1}{4\pi\left(1.0\times10^{-8}\,\text{s}\right)}$$

$$= 7.96\times10^{6}\,\text{s}^{-1}$$

$$= 8.0\times10^{6}\,\text{s}^{-1}$$

Since $\lambda = c/\nu$, then $|\Delta\lambda| = \dfrac{c}{\nu^2}|\Delta\nu| = \lambda\dfrac{|\Delta\nu|}{\nu}$.

With $\lambda = 610$ nm, $\nu = \left(3.00\times10^{8}\,\text{m s}^{-1}\right)/\left(610\times10^{-9}\,\text{m}\right) = 4.92\times10^{14}\,\text{s}^{-1}$, and

$$\Delta\lambda = \lambda\frac{\Delta\nu}{\nu}$$

$$= (610\,\text{nm})\left(\frac{7.96\times10^{6}\,\text{s}^{-1}}{4.92\times10^{14}\,\text{s}^{-1}}\right)$$

$$= 9.9\times10^{-6}\,\text{nm}$$

14.8   The resolution of visible and UV spectra can usually be improved by recording the spectra at low temperatures. Why does this procedure work?

The lower temperature reduces molecular speeds so that the effects of both Doppler and collisional broadening are reduced.

14.10   What is the molar absorptivity of a solute that absorbs 86% of a certain wavelength of light when the beam passes through a 1.0-cm cell containing a 0.16 M solution?

With 86% of light absorbed, the transmittance is $T = 1.00 - 0.86 = 0.14$, and the absorbance is $A = -\log T = -\log 0.14 = 0.854$. Then using the Beer–Lambert law, $A = \epsilon bc$,

$$\epsilon = \frac{A}{bc}$$

$$= \frac{0.854}{(1.0\,\text{cm})\,(0.16\,M)}$$

$$= 5.3\,\text{L mol}^{-1}\,\text{cm}^{-1}$$

14.12   A single NMR scan of a dilute sample exhibits a signal-to-noise (S/N) ratio of 1.8. If each scan takes 8.0 minutes, calculate the minimum time required to generate a spectrum with a S/N ratio of 20.

After n scans the signal intensity will increase by a factor of n, while the noise will increase by a factor of $\sqrt{n}$, so that, using $\left(\dfrac{S}{N}\right)_1$ for the signal-to-noise ratio after one scan, the ratio after n scans becomes

$$\left(\frac{S}{N}\right)_n = \frac{nS}{\sqrt{n}N} = \sqrt{n}\left(\frac{S}{N}\right)_1$$

or

$$\sqrt{n} = \frac{(S/N)_n}{(S/N)_1} = \frac{20}{1.8} = 11.1$$

$$n = 123$$

Acquiring 123 scans at 8.0 minutes per scan will require 123×8 min $= 9.8 \times 10^2$ min $= 16$ h.

14.14   What is the degeneracy of the rotational energy level with $J = 7$ for a diatomic rigid rotor? [The degeneracy is given by $(2J + 1)$.]

The degeneracy of a rotational energy level is given by $2J + 1$, so that for $J = 7$ the degeneracy is $2(7) + 1 = 15$.

14.16   The equilibrium bond length in nitric oxide ($^{14}N^{16}O$) is 1.15 Å. Calculate **(a)** the moment of inertia of NO, and **(b)** the energy for the $J = 0 \to 1$ transition. How many times does the molecule rotate per second at the $J = 1$ level?

In finding the reduced mass of $^{14}N^{16}O$ it is important to use masses appropriate for the specific isotopes under consideration and not the average masses found in the periodic table.

$$\mu = \frac{m_N m_O}{m_N + m_O} = \frac{(14.00\ \text{amu})\ (15.99\ \text{amu})}{14.00\ \text{amu} + 15.99\ \text{amu}} \left(1.661 \times 10^{-27}\ \text{kg amu}^{-1}\right) = 1.2399 \times 10^{-26}\ \text{kg}$$

(a) The moment of inertia is given by

$$
\begin{aligned}
I &= \mu r^2 \\
&= \left(1.2399 \times 10^{-26}\ \text{kg}\right) \left(1.15 \times 10^{-10}\ \text{m}\right)^2 \\
&= 1.640 \times 10^{-46}\ \text{kg m}^2 \\
&= 1.64 \times 10^{-46}\ \text{kg m}^2
\end{aligned}
$$

(b) The rotational constant for the molecule is

$$B = \frac{h}{8\pi^2 I} = \frac{6.626 \times 10^{-34}\ \text{J s}}{8\pi^2 \left(1.640 \times 10^{-46}\ \text{kg m}^2\right)} = 5.117 \times 10^{10}\ \text{s}^{-1}$$

and the energy for the $J = 0 \to 1$ transition is

$$\Delta E_{0 \to 1} = 2BhJ'$$

$$= 2 \left(5.117 \times 10^{10}\ \text{s}^{-1}\right) \left(6.626 \times 10^{-34}\ \text{J s}\right) (1) = 6.781 \times 10^{-23}\ \text{J} = 6.78 \times 10^{-23}\ \text{J}$$

The frequency of molecular rotation is equal to the frequency of the electromagnetic radiation that causes the transition, which is

$$\nu = \frac{\Delta E_{0 \to 1}}{h} = \frac{6.781 \times 10^{-23}\,\text{J}}{6.626 \times 10^{-34}\,\text{J s}} = 1.02 \times 10^{11}\,\text{s}^{-1}$$

14.18   Give the number of normal vibrational modes of **(a)** O_3, **(b)** C_2H_2, **(c)** CBr_4, **(d)** C_6H_6.

Molecules **(a)**, **(c)**, and **(d)** are non-linear and have $3N - 6$ normal modes, where N is the number of atoms in the molecule. Molecule **(b)** is linear and has $3N - 5$ normal modes.

(a) $3 \times 3 - 6 = 3$

(b) $3 \times 4 - 5 = 7$

(c) $3 \times 5 - 6 = 9$

(d) $3 \times 12 - 6 = 30$

14.20   A 500-g object suspended from the end of a rubber band has a vibrational frequency of 4.2 Hz. Calculate the force constant of the rubber band.

The vibrational frequency is given by $\nu = \dfrac{1}{2\pi}\sqrt{\dfrac{k}{m}}$, so that

$$k = 4\pi^2 m \nu^2$$
$$= 4\pi^2 \left(500 \times 10^{-3}\,\text{kg}\right)\left(4.2\,\text{s}^{-1}\right)^2$$
$$= 3.5 \times 10^2\,\text{kg s}^{-2}$$
$$= 3.5 \times 10^2\,\text{N m}^{-1}$$

14.22   If molecules did not possess zero-point energy, would they be able to undergo the $v = 0 \to 1$ transition?

Yes, since the oscillating electric field of the IR radiation would be able to induce motion of the centers of positive and negative charge in much the same way that the rotational motion of a polar molecule is excited. Of course, it is impossible to construct physically meaningful wavefunctions for an oscillating molecule with no zero-pont energy so that it might be argued that since such a molecule could never exist, neither could the transition take place.

14.24   Show all the fundamental vibration modes of **(a)** carbon disulfide (CS_2), and **(b)** carbonyl sulfide (OCS), and indicate which ones are IR active.

(a) The fundamental vibration modes of CS_2 are identical in form to those of CO_2 (see Figure 14.14). The asymmetric stretch and the (doubly degenerate) bending modes are IR active.

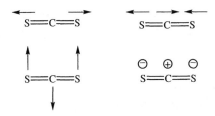

(b) OCS is also a linear molecule with 4 fundamental vibration modes. All four are IR active, leading to 3 IR peaks, since the bend is doubly degenerate.

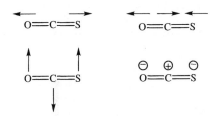

14.26 Which of the following molecules has the highest fundamental frequency of vibration? H_2, D_2, HD.

Since the fundamental frequency of vibration is given by $\nu = \dfrac{1}{2\pi}\sqrt{\dfrac{k}{\mu}}$, and since the force constant for the bond is isotopically invariant, the molecule with the lowest reduced mass will have the highest fundamental frequency of vibration. The reduced masses are

$$\mu_{H_2} = \frac{(1.008\ \text{amu})\,(1.008\ \text{amu})}{1.008\ \text{amu} + 1.008\ \text{amu}} = 0.5040\ \text{amu}$$

$$\mu_{HD} = \frac{(1.008\ \text{amu})\,(2.014\ \text{amu})}{1.008\ \text{amu} + 2.014\ \text{amu}} = 0.6718\ \text{amu}$$

$$\mu_{D_2} = \frac{(2.014\ \text{amu})\,(2.014\ \text{amu})}{2.014\ \text{amu} + 2.014\ \text{amu}} = 1.007\ \text{amu}$$

Thus, H_2 has the highest fundamental vibration frequency.

14.28 Anthracene is colorless, but tetracene is light orange. Explain.

Anthracene   Tetracene

In tetracene, the electrons are delocalized over a greater space than those in anthracene. Recalling the particle-in-a-box model, the greater the length of the "box", the smaller the spacing between energy levels. The greater "length" of the box in tetracene causes the absorption wavelength to shift from the UV region (in anthracene) into the visible region.

14.30 Many aromatic hydrocarbons are colorless, but their anion and cation radicals are often strongly colored. Give a qualitative explanation for this phenomenon. (*Hint*: Consider only the π molecular orbitals.)

In aromatic hydrocarbons, the HOMO is the highest energy π orbital and the LUMO is the lowest energy π* orbital. The energy separation between these orbitals is so large that the lowest energy transition, the $\pi \rightarrow \pi^*$, lies in the UV, rendering the molecule colorless. In the cation and anion radicals, the lowest energy transition is a $\pi \rightarrow \pi$ or $\pi^* \rightarrow \pi^*$ transition, respectively. The smaller energy spacings among the π and π* orbitals, compared to the HOMO-LUMO gap, result in the absorption of a photon in the visible region of the spectrum.

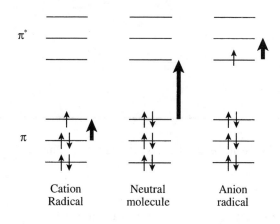

14.32 The NMR signal of a compound is found to be 240 Hz downfield from the TMS peak using a spectrometer operating at 60 MHz. Calculate its chemical shift in ppm relative to TMS.

According to Equation 14.50,

$$\delta = \frac{\nu - \nu_{ref}}{\nu_{spec}} \times 10^6 = \frac{240 \text{ Hz}}{60 \times 10^6 \text{ Hz}} \times 10^6 = 4.0 \text{ ppm}$$

14.34 What is the field strength (in tesla) needed to generate a ^{1}H frequency of 600 MHz?

From Equation 14.47 and Table 14.5,

$$B_0 = \frac{2\pi\nu}{\gamma} = \frac{2\pi \left(600 \times 10^6 \text{ s}^{-1}\right)}{26.75 \times 10^7 \text{ T}^{-1}\text{s}^{-1}} = 14.1 \text{ T}$$

14.36 For an applied field of 9.4 T (used in a 400-MHz spectrometer), calculate the difference in frequencies for two protons whose δ values differ by 2.5.

From Equation 14.50

$$\Delta\nu = \frac{\Delta\delta \times \nu_{spec}}{10^6} = \frac{2.5 \left(400 \times 10^6 \text{ Hz}\right)}{10^6} = 1.0 \times 10^3 \text{ Hz}$$

Values of Some Fundamental Constants

Constant	Value
Avogadro's constant (N_A)	6.0221367×10^{23} mol^{-1}
Bohr radius (a_o)	$5.29177249 \times 10^{-11}$ m
Boltzmann constant (k_B)	1.380658×10^{-23} J K^{-1}
Electron charge (e)	1.602177×10^{-19} C
Electron mass (m_e)	$9.1093897 \times 10^{-31}$ kg
Faraday constant (F)	96485.309 C mol^{-1}
Gas constant (R)	8.314510 J K^{-1} mol^{-1}
Neutron mass (m_N)	1.674928×10^{-27} kg
Permittivity of vacuum (ε_0)	8.854×10^{-12} C^2 N^{-1} m^{-2}
Planck constant (h)	6.626075×10^{-34} J s
Proton mass (m_P)	1.672623×10^{-27} kg
Rydberg constant (R_H)	109737.31534 cm^{-1}
Speed of light in vacuum (c)	299792458 m s^{-1}

Pressure of Water Vapor at Various Temperatures

Temperature/°C	Water Vapor Pressure/mmHg
0	4.58
5	6.54
10	9.21
15	12.79
20	17.54
25	23.76
30	31.82
35	42.18
40	55.32
45	71.88
50	92.51
55	118.04
60	149.38
65	187.54
70	233.7
75	289.1
80	355.1
85	433.6
90	525.76
95	633.90
100	760.00